SCIENCE AND COMMON SENSE

SCIENCE AND COMMON SENSE

BY JAMES B. CONANT

PRESIDENT OF HARVARD UNIVERSITY

NEW HAVEN YALE UNIVERSITY PRESS : 1951

LONDON : GEOFFREY CUMBERLEGE : OXFORD UNIVERSITY PRESS

In 1946 I had the privilege of delivering the Terry Lectures at Yale University. In these lectures I examined the question of how we can in our colleges give a better understanding of science to those who have no intention of being scientists and whose collegiate exposure to a natural science at best represents but a course or two. My proposed solution of a simple but difficult pedagogic problem can be summed up by saying that I advocated teaching the principles of the "tactics and strategy of science" by means of a series of "case histories." To illustrate what I had in mind it was necessary to outline a few sample cases —one dealing with Robert Boyle's work in pneumatics, one dealing with Galvani's and Volta's discoveries in electricity, and one outlining the chemical revolution associated with the name of Lavoisier. These sketches of what might become case histories, together with my own interpretation of the tactics and strategy of science, were presented to the public in a small volume entitled *On Understanding Science, An Historical Approach.*

When the question of preparing a second edition of this

book arose some twelve months ago, it was clear that something more than a revision was required. *On Understanding Science* had attempted a twofold task; I had endeavored to give the general reader some understanding of the methods of science and to outline for the college teacher how one type of instruction might possibly be carried on. Any further discussions of my proposals for teaching science to nonscientists would now have to consider what has happened in the last five years in this country in the teaching of college science; reports of progress would require a discussion of pedagogic details of little interest to those outside the teaching profession. On the other hand, three years' experience with an undergraduate course in Harvard College as part of the program of presenting Natural Science within the framework of General Education had both somewhat altered my views about scientific tactics and provided me with copious illustrative material.

Therefore, rather than revise and amplify a small book addressed to two separate audiences, I decided to attempt a much larger book on the methods of science for the general reader. In so doing I have kept those portions of *On Understanding Science* which were relevant to the new orientation of my discussion of the tactics and strategy of science. References to the teaching of science have been all but eliminated. Any who are curious as to the development of a freshman course devoted to giving an understanding of science and scientists with the help of case histories are referred to the Harvard Case Histories in Experimental Science. The present volume, while paralleling to some degree the Harvard course with which I have been associated, does not give a complete picture of what has been there attempted. Nevertheless, I am hopeful that it will be quite adequate to provide a briefing for the busy citizen as to the way scientists carry on their work.

Exposition for the general reader is, of course, something quite different from the presentation of the same point of view to a college class. The old saying that "Education is what is left after all that has been learnt is forgotten" implies a stiff process of study without which college courses would be as ephemeral in their influence as lectures to women's clubs. But the general reader must be approached in a different way. He can be given only an outline; if he has time and the inclination he may fill in the gaps through his own explorations, by conversation with experts, or through the printed word. A short bibliography is, therefore, appended for any who wish to study more exhaustively examples of "science in the making" or who are interested in exploring further certain of the topics connected with science and society.

Except for Chapter X ("The Study of the Past"), essentially all the ideas presented in the following pages and much of the technical material have been presented by the author more than once to audiences of undergraduates. Needless to say, as a consequence of this experience both the ideas and the mode of presentation have been more than once subject to drastic change. This book, therefore, reflects the impact of the student upon the teacher. To a far greater degree it reflects the impact of a group of teachers on the author. For during the better part of three academic years I have had the opportunity of a weekly luncheon discussion with those who have collaborated with me in the Natural Science course to which I have referred. From the give and take of these meetings has evolved almost all that is new in this book as compared with the Terry Lectures. It is obviously impossible to make acknowledgment of the specific contributions of the various members of the group. From the robust nature of our discussions I feel certain that each one of them will find much in my treatment of *Science and Common Sense* with which he does not agree. I feel equally sure there will be little unanimity on the points chosen for disagreement. But as to my great indebtedness to them collectively and individually there is in my mind not the slightest doubt. Therefore I take this opportunity of once again expressing my gratitude to Professor Gwilym E. Owen of Antioch College, Professor Duane Roller of Wabash College, Associate Professor Fletcher G. Watson of Harvard, Assistant Professor Leonard K. Nash of Harvard, Dr. Thomas S. Kuhn of Harvard, Dr. Charles L. Clark, and Dr. Eugene P. Gross.

The whole enterprise would never have been undertaken four years ago if I had not been able to obtain the support of a learned young historian of science, Assistant Professor I. Bernard Cohen. He has continued to lend a helping hand during the past three years. To Mr. Duane H. D. Roller I am indebted for assistance in the preparation of the manuscript of the first nine chapters and the illustrations. Last but by no means least, I wish to express my thanks to my wife, Grace Richards Conant, for help with both the manuscript and the proof.

JAMES BRYANT CONANT

Cambridge, Massachusetts
October 12, 1950

CONTENTS

Contents

CHAPTER ONE

Science and the American Citizen

THIS IS an elementary book about the methods of experimental science. The reader is assumed to have an interest in the procedures of the scientist in his laboratory but only a bowing acquaintance with the principles of the natural sciences. The object is to provide him with some understanding of the way in which physicists, chemists, biochemists, and experimental biologists tackle their problems and some knowledge of how their efforts are related to progress in technology, agriculture, and medicine.

In other words, this is a citizen's guide to the methods of experimental science. The exposition that follows is addressed to the intelligent citizen who as a voter may, to an increasing extent, be interested in congressional action on scientific matters. I have likewise in mind the lawyer, banker, industrialist, government official, politician, or newspaperman who is deeply immersed in the practical problems of the mid-twentieth century. On every side he or she will encounter scientists at work or the consequences of their work. In industry, in the hospitals, in the experiment stations, in the universities, questions arise every year which involve the use of the words "research" or "development" or "scientific investigation"; they generally also involve two less attractive combinations of letters, "budget" and "costs." How can a man without scientific training know whether the enthusiastic chemist, doctor, or engineer is talking sense when he urges the investment of money in this or that adventure? This query is relevant to decisions made almost daily by boards of directors of all manner of business undertakings, by trustees of hospitals and universities, by federal and state officials, and by members of Congress.

No magic pill can be administered to make a person capable of matching wits with an expert. But it is possible to acquire through long experience some appreciation of the point of view of the laboratory scientist, some understanding of his methods and the way he conceives a problem. Not a few American citizens have made themselves intelligent critics of scientific work by a process of self-education in relation to some phase of modern science connected with their activities in the world of affairs. A series of episodes has left permanent imprints on their minds which serve as fixed points—a sort of map to which they may refer any new proposal from the laboratory.

Those who are just starting their professional careers may find the exposition of the methods of science set forth in the following pages of some value. Others who have great practical experience but have had no working relationship with scientists or engineers may desire some glimpse of the ways of modern science; they may hope for an introduction to the understanding which has come to a few laymen through long years of special responsibility.

And every American citizen, if he is sufficiently youthful and ambitious, is a potential leader of his community and may ultimately be responsible for decisions and expenditures. In public health and medicine, or in business as a manager or labor leader, he may be hard pressed in determining policy involving science or its applications; or just as a plain man on the street, a voter, he may have to register his opinion on matters of high national import which concern research and development projects financed by public funds. Of such projects in the tragic days of the 1950's none seems of more immediate significance than the development of weapons. Whether we like it or not, we are all immersed in an age in which the products of scientific inquiries confront us at every turn. We may hate them, shudder at the thought of them, embrace them when they bring relief from pain or snatch from death a person whom we love, but the one thing no one can do is banish them. Therefore every American citizen in the second half of this century would be well advised to try to understand both science and the scientists as best he can.

One may at this point inquire what I have in mind when I speak of a layman's understanding of science. Let me explain. In my experience, a man who has been a successful investigator in any field of experimental science approaches a problem in pure or applied science, even in an area in which he is quite ignorant, with a special point of view. We may designate this point of view "understanding science." Note carefully that it is independent of a knowledge of the scientific facts or techniques in the new area to which he comes. Even a highly educated and intelligent citizen without research experience will almost always fail to grasp the essentials in a discussion which takes place among scientists concerned with a projected inquiry. This will be so not because of the layman's lack of scientific knowledge or his failure to comprehend the technical jargon of the scientist; it will be to a large degree because of his fundamental ignorance of what science can or cannot accomplish and his consequent bewilderment in the course of a discussion outlining a plan for a future investigation. He has no "feel" for the tactics and strategy of science.

In the last ten years I have seen repeated examples of such bewilderment of laymen. If I am right in this diagnosis (and it is the fundamental premise of this book), the remedy does not lie in a greater dissemination of scientific information among nonscientists. Being well informed about science is not the same thing as understanding science, though the two propositions are not antithetical. What is needed is methods for imparting some knowledge of the tactics and strategy of science to those who are not scientists. Not that one can hope by any short-cut methods to produce in the mind of a nonscientist the same instinctive reaction toward scientific problems that is the hallmark of an investigator; but enough can be accomplished, I believe, to bridge the gap to some degree between those who understand science because science is their profession and intelligent citizens who have only studied the results of scientific inquiry—in short, the laymen.

One way to proceed would be to require every person to take a few years off and become a visitor to scientific institutions. He might spend six months, let us say, at the elbow of a group leader in a research laboratory of some large chemical company; then move to a corresponding vantage point in the electrical industry, then to a university laboratory of physics or chemistry, finally to a hospital or to the engineering operations of those who are studying new ways of utilizing coal. Any number of alternative plans for direct observation of scientists at work could be easily drawn. The detailed specifications might vary; it could be argued that most of the time should be spent in universities or, on the other hand, that industrial laboratories would be more relevant, but we can all agree that by a number of such hypothetical operations a person totally ignorant of even high-school physics, chemistry, or biology could in the course of a few years acquire a good understanding of science and its methods.

Unfortunately, there are grave difficulties quite apart from the time involved: visitors are proverbial nuisances in any laboratory; also, there are long periods when nothing really interesting or exciting happens. Furthermore, we should have to provide some mechanism for having the visitors arrive only on the crucial days and ask only highly intelligent questions which the scientist would be not too busy to answer patiently. And we should really provide some magic by which the whole visit could be repeated for those who failed to respond to the first exposure, just as we show an educational film more than once to some student audiences!

But I have overlabored this bit of fantasy. To anyone who has scanned the chapter headings the point I am trying to make is obvious. By taking the reader back to certain events in scientific history, I am proposing to do the equivalent of the magic tour of laboratories that has just been described. The word "equivalent" is too strong; rather, one might say that something of the experience which would come from visiting various laboratories may be communicated by a discussion of the methods by which scientists have advanced knowledge in the past. For the amount of time spent, at least, a comparable degree of understanding may result. And if someone objects (as many already have) that my illustrations are all taken from those periods when a given branch of science was in its infancy, that I am offering history to those hungry for current information, I must reply, the methods of science were the same then as now and only by this procedure can a sufficiently simplified exposition be provided.

It is not the present moment which is of interest to the scientist or to those who look over his shoulder (or look him in the eye when he asks for funds). It is the future. This point will be emphasized throughout this book. Present scientific knowledge is of interest only to encyclopedists. If all research, all effort to find new ways of doing things were suddenly to cease, the citizen as well as the scientist would lose interest in the whole affair. The significance of science in our time is that something is happening every day in countless laboratories, pilot plants, and hospitals. These new things are happening because certain complex patterns of human behavior have been developed over the last three hundred years. These we may call the methods of experimental science. To disentangle the many threads in the present pattern in any one subdivision of the highly complex structure of modern science is so difficult as to be almost a hopeless task—certainly impossible to accomplish through the medium of one book. But by studying significant advances at an early stage in the history of a given field, we can to some degree avoid the confusion caused by the complexities of modern science. In so doing we run the danger, of course, of forgetting that science today is a closely woven fabric, that multitudinous threads with long and diverse histories mutually support one another. From time to time we will remind ourselves of this situation in the 1950's by referring ahead from the simple illustration of an earlier age to the days in which we live.

THE TRADITIONS OF SCIENTIFIC INQUIRY

One of the most significant discussions now in progress turns on how far the methods by which the astonishing results in the physical and biological sciences have been achieved may be transferred to other human activities. Among the questions on which learned and sincere men now disagree are the following: Is there such a thing as a scientific method of wide applicability in the solution of human problems? Are the social sciences really sciences?

The answers to these and many related questions are of importance to the future of a free people. Our educational procedures on the one hand and our collective actions in regard to a variety of social, economic, and political problems on the other may well depend on our assessment of the future of the social sciences. Now, obviously, if a layman is to have any clear ideas about the relation of the methods of physics, chemistry, and biology to education or to the investigation of human problems, he must understand the methods in question. There seems to be a profound need for clarification of popular thinking about the methods of the natural sciences. This is required in order to lay the basis for a better discussion of the ways in which rational methods may be applied to the study and solution of all manner of human problems.

One extreme position which has been maintained with some insistence for many years attempts to equate the scientific method with all relatively impartial and rational inquiries. Somewhat more than sixty years ago, for example, Karl Pearson in his *The Grammar of Science* proclaimed that "Modern science, as training the mind to an exact and impartial analysis of facts, is an education specially fitted to promote sound citizenship." And for the busy layman he recommended that "What is necessary is the thorough knowledge of some small group of facts, the recognition of their relationship to each other, and of the formulae or laws which express scientifically their sequences. It is in this manner that the mind becomes imbued with the scientific method and freed from individual bias in the formation of its judgments—one of the conditions, as we have seen, for ideally good citizenship. This first claim of scientific training, its education in method, is to my mind the most powerful claim it has to state support."

With Pearson's statement about the scientific method I shall have occasion to quarrel later. For the moment let us concentrate attention on two implications that run through the first portion of his book: first, that an exact and impartial analysis of facts is alone possible in the realm of science; second, that exposure to such a discipline will produce a frame of mind that makes for impartial analysis in all matters.

Now there is no question that one of the necessary conditions for scientific investigation is an exact and impartial analysis of the facts. But this attitude was not invented by those who first concerned themselves with scientific inquiries nor was its overriding importance at once recognized. As one skims the histories of the natural sciences, it seems clear that in the embryonic stages of each of the modern disciplines, violent polemics rather than reasoned opinions often flowed most easily from the pen. If I read the history of science in the seventeenth and eighteenth centuries rightly, it was only gradually that there evolved the idea that a scientific investigator must impose on himself a rigorous self-discipline the moment he enters his laboratory. As each new generation saw how the prejudice and vanity of its predecessors proved stumbling blocks to progress, standards of exactness and impartiality were gradually raised. But as long as science was largely a field for amateurs, as it remained into the nineteenth century, a man could regard his discoveries like so many fish. If he defended their size against all detractors, and if in the process their length increased, well, his opponent was a well-known liar, too!

The formation of the scientific societies, their growing importance, and the gradual building up of a professional feeling about science slowly changed the atmosphere. The example of the few giants, exemplified by Galileo, who had recognized the need for self-control became the accepted standard. The man who was inclined to use the same weapons in "philosophical" as in political and theological debate gave way to the modern scientist who places little reliance on persuading his opponent with rhetoric or driving him from the field with invective. For his jury today is a large body of well-informed peers and to them he need only present accurate reports with the minimum of emotion. I am referring to scientists who speak to scientists, of course. I leave aside the great popularizers of science like Huxley who are really educators.

Would it be too much to say that in the natural sciences today the given sociological environment has made it very easy for even an emotionally unstable person to be exact and impartial in his laboratory? The traditions he inherits, his instruments, the high degree of specialization, the crowd of witnesses that surrounds him, so to speak (if he publishes his results)—these all exert pressures that make impartiality on matters of *his* science almost automatic. Let him deviate from the rigorous role of impartial experimenter or observer at his peril; he knows all too well what a fool So-and-so made of himself by blindly sticking to a set of observations or a theory now clearly recognized as in error. But once he leaves the laboratory behind him he can indulge his fancy all he pleases and perhaps with all the less restraint because he is now free from the imposed discipline of his calling. One would not be surprised therefore if, as regards matters beyond their professional competence, laboratory workers were a little less impartial and self-restrained than other men, though my own observations lead me to conclude that as human beings scientific investigators are statistically distributed over the whole spectrum of human folly and wisdom much as other men.

Who were the precursors of those early investigators who in the sixteenth and seventeenth centuries set the standards for exact and impartial scientific inquiries? Who were the spiritual ancestors of Copernicus, Galileo, and Vesalius? Not the casual experimenters or the artful contrivers of new mechanical devices who gradually increased our empirical knowledge during the Middle Ages. These men passed on to subsequent generations many facts and valuable methods of attaining practical ends but not the spirit of scientific investigation. For the burst of new ardor in disciplined intellectual inquiry we must turn to a few minds steeped in the Socratic tradition and to the early scholars who first recaptured the culture of Greece and Rome by primitive methods of archaeology. In the first period of the Renaissance the love of dispassionate search for the truth was carried forward by men who were concerned with man and his works rather than with inanimate or animate nature. During the Middle Ages interest in attempts to use the human reason critically and without prejudice, to probe deeply without fear and favor, was kept alive by those who wrote about human problems.

In the early days of the revival of learning it was the humanist's exploration of antiquity that came nearest to exemplifying our modern ideas of impartial inquiry. Until the wave of scientific curiosity began to mount, inquiries into what we now call natural science were of little interest even to educated men. Scientific conclusions, unless they had profound influence on current cosmology, were apt to be lost like a pebble tossed into the sea.

Why and how the wave of scientific curiosity began to mount are, of course, among the most fascinating and difficult of historical questions. There are no simple answers. Any exposition of the "dawn of science" is certain to be in error by emphasizing unduly either one or another of a group of complex factors that were clearly at work in shaping the new era in which we live. I have heard an eminent historian of the culture of the Middle Ages declare that the humanists contributed nothing whatsoever to the beginning of modern science; in fact, their activities were probably detrimental. On the other hand, any assertion that the recapture of the writings and spirit of antiquity by the humanists was the sole determining factor in the development of science is surely an overstatement.

Galileo admittedly obtained inspiration and ideas as a young man from reading Archimedes; from this one can argue that the role of the revival of learning in stimulating science was great. A Latin translation of the Greek would not have come into the hands of a man of Galileo's bent and genius three hundred years earlier, for the first Latin translation of Archimedes was made by William of Moerbeke and was printed in 1543. The publication of a Latin translation of the work of Hero of Alexandria in 1575 served to arouse interest in hydraulic phenomena.

But more important than contact with the ancient world

or even the increasing spread of information due to the invention of printing was the spirit of intellectual adventure so characteristic of the Italian city republics in their days of glory. The story told of Filippo Brunelleschi by Vasari has always seemed to me to epitomize the boiling up of the curiosity and creative energy of the Renaissance which eventually manifested itself in scientific investigation. It is in brief as follows: "One morning, some months after his return, Filippo was on the piazza of S. Maria del Fiore with Donato and other artists discussing antique sculptures, and Donato was relating how . . . he had made a journey to Orvieto . . . and how, in passing afterwards through Cortona, he . . . had seen a remarkable ancient marble sarcophagus, with a bas-relief, a rare thing then, . . . and so inflamed Filippo with an ardent desire to see it that, just as he was, in his mantle, hood and sabots, he left them without saying a word . . . and proceeded to Cortona led by his love and affection for art."

In this same connection Charles Singer writes in his *Short History of Biology,* "The beginnings of effective plant study have been traced to a fortunate combination of Humanistic Learning, Renaissance Art, and the perfection of the Craft of Printing. The same is true of the study of the animal body."

Science became a self-propagating social phenomenon, according to my view, when the ferment of the Italian Renaissance underwent a mutation and then spread through new generations of young men; the focus of attention turned from art, archaeology, and literature to the study of the structure of plants and animals, the stars, and to mechanical contrivances. If one may carry the metaphor further, this new strain of ferment was able to find a lodging in what had hitherto proved a barren medium. People less sensitive to poetry and art than the inhabitants of Italian cities and towns could share the enthusiasm of those who made discoveries about the human body, or stars, or falling bodies, or ways of creating vacuums. Galileo seems to have been a man quite in the spirit of Brunelleschi; Boyle and his friends in the Oxford of the 1650's (of whom I speak in a later chapter) would have had much in common with Galileo, but I cannot imagine them transported in time and place to become boon companions of Brunelleschi. They were too near Milton in several senses of the word.

If the preceding analysis of history be correct, Petrarch, Boccaccio, Machiavelli, and Erasmus, far more than the alchemists, must be considered the spiritual precursors of the modern scientific investigator. Likewise, Rabelais and Montaigne who carried forward the critical philosophic spirit must be counted among the forerunners of the modern scientists. Not only the Renaissance antiquarians, and a few hardy skeptics, but also honest explorers and hardheaded statesmen were the ancestors of all who have since endeavored to find new answers to old questions, who desire to minimize prejudice and examine facts impartially. As I see it, scientists today represent the progeny of one line of descent who migrated, so to speak, some centuries ago into certain fields which were ripe for cultivation. Once science had become self-propagating, those who till these fields have had a relatively easy time keeping up the tradition of their forebears.

Therefore, to put the scientist on a pedestal because he is an impartial inquirer is to misunderstand the situation entirely. Rather, if we seek to spread more widely among the population the desire to seek facts without prejudice, we should pick our modern examples from the nonscientific fields. We should examine and admire the conduct of the relatively few who in the midst of human affairs can courageously, honestly, and intelligently come to conclusions based on reason without regard for their own or other people's loyalties and interests and, having come to these conclusions, can state them fairly, stick by them, and act accordingly.

To say that all impartial and accurate analyses of facts are examples of the scientific method is to add confusion beyond measure to the problems of understanding science. To claim that the study of science is the best education for young men who aspire to become impartial analysts of human affairs is to put forward a very dubious educational hypothesis at best. Indeed, those who contend that the habits of thought and point of view of the scientist as a scientist can be transferred with advantage to other human activities have hard work documenting their proposition.

With idolatry of science I must confess I have little sympathy. Yet a better understanding of the methods of the natural sciences among laymen is certainly to be desired. Since scientific investigations provide widespread and often dramatic examples of an effective way of handling problems, a greater knowledge of the genesis of scientific methods ought to be imparted by our schools and colleges. Today the artificial restraints under which the experimentalist now unconsciously operates make cold-blooded factual analysis almost a routine operation, and the demonstration day after day of the success of such methods has profoundly influenced public opinion. To be properly understood, this demonstration, which strengthens other rational elements in our civic life, must be seen as the consequence of a sociological process whose history we can

trace for at least three centuries. In addition to understanding something about the methods of science, a citizen needs to understand the operations of science as a
human enterprise.

SCIENCE AS AN ORGANIZED ACTIVITY

The physical and biological sciences today consist of a
closely interlocking set of principles and theories and a
vast amount of classified information. They are also the
product of a living organization. The theories, laws, data
are to be found in libraries, herbariums, and museums;
these are useful residues, deposits of the past, but essentially dead material. The activity we associate with the
word "science" is the sum total of the potential findings
of the workers in the laboratories; it is their plans, hopes,
ambitions in the process of realization, week after week,
year after year, that is the essence of modern science. Now
this is a clear case, if there ever was one, of the whole being
something quite different from the sum of the parts. For
if the thousands of experimental scientists who are going
to their laboratories tomorrow were not able to communicate with each other rapidly and easily there would be
no modern science.

This is something far more complicated and far more
important than the layman often realizes. Indeed, to a misunderstanding of the nature of science as an organized
social activity today one can trace many a foolish statement and some practical blunders. It is amazing how much
credence is given to self-deluded quacks or real charla-
tans, or how old wives' tales become accepted as scientific
statements. The tendency to equate science with magic
can be seen on almost every hand. This man tells you in all
seriousness that he knows someone who knows how to stop

an automobile engine at a distance of a mile by whistling the proper note; another believes that an untrained amateur has discovered a way of making real, natural rubber directly from garbage in one step; not to mention the whole host of bogus and pseudoscientific "cures" and remedies which still invade the field of medicine.

No one can be blamed for not detecting an absurd statement about alleged principles in physics, chemistry, or biology. All of us who have been engaged for many years in teaching these subjects or writing textbooks in these fields have had the experience of having to revise radically some of our basic statements to keep abreast of the advance of knowledge. But the first reaction of almost any scientist to a rumor about an alleged new step forward is one of incredulity. Maybe the step is a false one—he recalls at once the number of such instances in the field of his own experience. However, he is quite sure that in a short time the matter will be settled, unless some very revolutionary step is in the making. By the process of "publication" the new idea or new experimental finding will be made available to a host of scientists all over the world. Before long others will subject the report to critical examination if the matter is of real importance. No startling or even arresting alleged discovery will remain unnoticed.

It is not a question merely of repeating a series of calculations or checking experimental findings. There will be hundreds of implications for workers in the same or allied fields of inquiry and these will be followed up. If they fail to yield the expected result the original report may be dismissed by the scientific world as "another pipe dream." Eventually the author of the now discredited paper may discover his own error and publish a correction, or the whole matter may simply be allowed to drop. One could write a large volume on the erroneous experimental findings in physics, chemistry, and biochemistry which have found their way to print in the last hundred years; and another whole volume would be required to record the abortive ideas, self-contradictory theories and generalizations recorded in the same period.

The important fact which emerges from even a superficial study of the recent history of the experimental sciences (say, since 1850) is the existence of an organization of individuals in close communication with each other. Because of the existence of this organization new ideas spread rapidly, discoveries breed more discoveries, and erroneous observations or illogical notions are on the whole soon corrected. The deep significance of the existence of this organization is often completely missed by those who talk about science but have had no firsthand experience with it. Indeed, a failure to appreciate how scientists pool their information and by so doing start a process of cross-fertilization in the realm of ideas has resulted in some strange proposals by politicians even in the United States. And in the Soviet Union we see what is apparently a deliberate attempt to alter drastically the nature of science as an organized social activity.

Science as a profession, we must remember, is a recent invention. Some of the most important advances in the early development of physics and chemistry were made by amateurs. Indeed, in the examples given in the following chapters to illustrate the methods of science we shall encounter very few men who earned their living by scientific investigation or even by teaching science. As a rough generalization one may say that modern science started in the Italian universities in the sixteenth century; it flourished in the same environment until about the middle of the seventeenth century, and then the focus of activity is to be found in Paris and in London. The role of the universities does not become significant again until the nineteenth century. The seventeenth and eighteenth centuries were the period of the learned societies, particularly the Royal Society of London and the Académie des Sciences of Paris.

The significance of the Royal Society and the French academy of science lies in the fact that these *formal* institutions are the points of origin from which have grown the informal but highly complex organizations of modern science. The Royal Society was chartered by Charles II just after the Restoration but can trace its origin to the enthusiasm of a group of amateur scientists whom the accidents of party politics placed in Oxford in the Cromwellian period (1650–60). The French academy was the creation of Louis XIV in 1666 acting on the advice of Colbert. The intellectual parent of both is generally stated to be Francis Bacon, for in his unfinished fable, "The New Atlantis," published in 1626 shortly after his death, this ardent proponent of "the new experimental philosophy" (which he only partially understood and never practiced) described a House of Salomon which was a community of investigators and philosophers. An actual society founded in Rome in 1600, the Accademia dei Lincei, seems to have been a prototype of the organization Bacon envisaged. Galileo was a member of this academy which, as early as 1612, was stated to be a "gathering . . . which directs its labors diligently and seriously to studies as yet little cultivated." A generation later a group of Galileo's disciples in Florence established the Accademia del Cimento (1657) which flourished for ten years under the patronage of the two Medici brothers, Grand Duke Ferdinand II and

Leopold, who had been pupils of Galileo. This Academy of Experiment was more nearly the forerunner of a twentieth-century research institute than of the eighteenth-century learned society, for the members were engaged in cooperative experimentation, of which more will be said in another chapter.

Both these Italian scientific societies were in the tradition of the literary clubs which flourished in the centers of Renaissance culture. The history of both the Royal Society and the French academy shows a certain ambiguity as between two goals; on the one hand, the members envisioned an active cooperating group of experimenters, and on the other merely a meeting place for reporting and discussing experimental findings, quaint observations, and new ideas reported on a highly individualistic basis. The fact that the Royal Society never had any more support from the government than a royal blessing prevented any serious attempt to be much more than a focal point for discussion. The French monarchy, on the other hand, made grants to some of the members of the academy, and off and on during the course of a century thus supported scientists as a royal patron might support painters and men of letters.

Some of the expeditions organized and financed by these societies are of significance in the history of science. But the prime importance of the scientific societies lies in the fact that each undertook to publish a regular journal in which members and others could record their ideas and experimental results. Of the *Transactions* of the Royal Society whose publication was started in 1665 Huxley once said, "If all the books in the world except the Philosophical Transactions were destroyed, it is safe to say that the foundations of physical science would remain unshaken,

and that the vast intellectual progress of the last two centuries would be largely, though incompletely, recorded."
(One may doubt if those concerned today with the more
descriptive sciences such as mineralogy and organic chemistry would quite agree with this remark made in the
nineteenth century.)

Before the founding of the scientific societies and the
establishment of regular quarterly or monthly magazines
devoted to publishing the results of original work, news of
scientific discoveries spread by letter. Then, from time to
time, a scientific investigator would publish a small book
giving his ideas and recounting his experiments. This practice of using separate books rather than communications
to journals to announce scientific findings continued late
into the nineteenth century. But the scientific journals became more and more important; books are now reserved
for the purpose of summing up or amplifying the results
published elsewhere. Today the scientific journals, not the
scientific books, are the sources of information about what
is going on among those who labor on the frontiers of
knowledge.

To the uninitiated it would seem impossible for anyone
to find his way through the mass of articles and reports
which fill tens of thousands of pages every year. Actually
the task, while time consuming, is far from hopeless for
an investigator who makes a practice of following what
the scientist calls "the current literature." In the first place,
it must be remembered that by the beginning of the twentieth century the division and subdivision of the sciences
had proceeded very far. The journals of the scientific societies still accepted communications covering a wide range
of subjects, but specialized journals sprang up as early as
the first half of the nineteenth century. Therefore, today

one can keep abreast of the advance of science in a particular field of inquiry by reading a relatively small fraction of the monthly output of learned papers. In the second place, various elaborate schemes of indexing and abstracting have been devised. In some branches of science vast encyclopedias are published which summarize the results obtained under appropriate headings. A student bent on becoming an investigator soon learns how "to use the literature." Third, the custom has been established of referring to previous publications which bear on the subject under study. Finally, a great mass of irrelevant data and poorly digested reporting is eliminated by the activities of the editors of the journals. There is some danger in this process. More than one historic instance stands as evidence of originality taking such unusual forms as to cause a too conservative group of editors to regard the communication as erroneous or so fantastic as to warrant rejection. But there are so many different journals today that at the most there is likely to be but a short delay before publication. Informed criticism would say that far too much is published which might well be refused rather than that editors were too severe in their judgments.

The whole question of recording inventions and discoveries of a practical nature is one which we shall postpone considering until, in the concluding chapter, the interrelation of pure and applied science is reviewed. At that time a word about patents and the patent literature will be in order. Here I wish to emphasize that the methods of communicating scientific news on a reliable basis have evolved today to a point where there is little danger that any significant new discovery will remain unnoticed. If science were magic, some secret recipe might remain in the possession of a single individual; he might demonstrate the wonders worked by its aid without revealing the nature of his knowledge. As late as the eighteenth century, when chemistry was working out from under the shadow of alchemy, some new experimental procedures were kept secret for periods of time. Except as regards the industrial application of science, these days are long past. Or so we all thought before 1940. The close relation of some new knowledge in atomic physics to the production of weapons makes it now necessary to introduce a highly disturbing reservation into the story of the growth of rapid and uncensored publication of scientific findings. Likewise, the policy of the Communist party on the other side of the Iron Curtain requires the addition of a provision that the preceding description of scientific communication applies in the 1950's only to those who work in the free nations of the world.

We shall return to a discussion of these two mid-twentieth-century anomalies. In a later chapter the changing status of the scientist over the centuries will be reviewed. Science and politics is a topic of current interest. The relation of science to society has been a favorite theme of those concerned with the public welfare. Few who discussed this topic in the early thirties could have seen the problems presented to both scientists and society by the development of the atomic bomb on the one hand and the hardening of the dictatorship of the Kremlin on the other. But for a fruitful discussion of all such questions some background of understanding science is required. Let us turn to a consideration of the methods of science and first of all try to answer the question: What is science?

CHAPTER TWO

What Is Science?

O NE COULD fill several pages with quotations defining
science. What the man in the street means by science, however, is fairly clear. He has in mind the activity
of people who work in laboratories and whose discoveries
have made possible modern industry and medicine. Those
who implicitly or directly denounce science may have in
the forefront of their thought the application of science to
war, particularly the use of atomic energy. Others, who to
promote a pet idea label it "scientific," seek to evoke a
favorable response based on the widespread knowledge
of the highly beneficial effects of the application of science, particularly in medicine. In short, when the noun
and the adjective are used as they often are to buttress an
argument, the overtones will vary with the preoccupation
of the speaker. Perhaps what follows may be regarded by
some as no exception to this rule. The primary purpose of
this book is to assist the layman in understanding what
goes on in a modern laboratory. He may then relate this

to other allied activities which may or may not be designated as scientific. But a preoccupation with the experimental sciences is surely justified even in a treatment of the broad subject, "Science and Common Sense." For no definition of science excludes physics, chemistry, and experimental biology, and no one can deny that it is the rapid progress made in these fields and the startling applications of the knowledge thus obtained which give science its place in modern civilization.

But limiting one's attention merely to the experimental sciences by no means provides a satisfactory answer to the question "What is science?" For, immediately, diversity of opinion appears as to the objectives and methods of even this restricted area of human activity. The diversity stems in part from real differences in judgment as to the nature of scientific work but more often from the desire of the writer or author to emphasize one or another aspect of the development of the physical and biological sciences. There is the static view of science and the dynamic. The static places in the center of the stage the present interconnected set of principles, laws, and theories, together with the vast array of systematized information: in other words, science is a way of explaining the universe in which we live. The proponent of this view exclaims "How marvelous it is that our knowledge is so great!" If we consider science solely as a fabric of knowledge, the world would still have all the cultural and practical benefits of modern science, even if all the laboratories were closed tomorrow. This fabric would be incomplete, of course, but for those who are impressed with the significance of science as "explanations" it would be remarkably satisfactory. How long it would remain so, however, is a question.

THE DYNAMIC VIEW OF SCIENCE

The dynamic view in contrast to the static regards science as an activity; thus, the present state of knowledge is of importance chiefly as a basis for further operations. From this point of view science would disappear completely if all the laboratories were closed; the theories, principles, and laws embalmed in the texts would be dogmas; for if all the laboratories were closed, all further investigation stopped, there could be no re-examination of any proposition. I have purposely overdrawn the picture. No one except in a highly argumentative mood would defend either the extreme static or the extreme dynamic interpretation of the natural sciences. But the presentation of elementary science in school and college and the popular accounts almost necessarily take the dogmatic form, and therefore the American citizen is apt to be unconsciously drawn much too far in the one direction. The worker in the laboratory, however, clearly would not be there if he were not primarily concerned with science as an exploration. To understand him and his predecessors who have advanced the sciences since the sixteenth century one can hardly overemphasize the dynamic nature of science.

At all events, this is my own prejudice and I shall make no attempt to conceal it. My definition of science is, therefore, somewhat as follows: Science is an interconnected series of concepts and conceptual schemes that have developed as a result of experimentation and observation and are fruitful of further experimentation and observations. In this definition the emphasis is on the word "fruitful." Science is a speculative enterprise. The validity of a new idea and the significance of a new experimental finding are to be measured by the consequences—consequences in terms of other ideas and other experiments. Thus conceived, science is not a quest for certainty; it is rather a quest which is successful only to the degree that it is continuous.

This last statement may seem to some on first reading as equating scientific activity with a form of madness. Why this frantic chase of concepts and conceptual schemes where validity depends only on their breeding new experiments which in turn generate new ideas and so on, ad infinitum? Unless you are going to seek to justify science solely in terms of its application (which I do not), is not this dynamic view a defeatist view? Why not boldly claim, as many scientists have in the past, that the physicists and the chemists are trying to find out how the inanimate universe is constructed and how it works? If that is the goal, then clearly there is a terminal point, at least in principle; when the puzzle has been solved, the structure of the universe discovered, the laboratories can be closed and mankind take up other tasks. This is common sense, you may say; idle talk about fruitful concepts and conceptual schemes diverts both the layman and the scientist from the "hard facts" on which science rests!

These problems, if treated adequately, would require a separate book written by a group of philosophers. I use the word "group" advisedly because there is no one accepted answer to some of the difficult questions which are implied. In an elementary exposition of scientific methods one may be permitted to by-pass to a large extent the worries of the learned men who wrestle with the problem of how human beings can know anything or indeed what the word "know" may mean. But since the common-sense idea of what scientists are about is usually vaguely formulated in terms of discovering how the universe is *really* constructed, some paragraphs to justify my cautious handling of scientific findings must be here inserted.

SCIENCE AND REALITY; A SKEPTICAL APPROACH

As the title of this book indicates, I am hopeful that by relating science to common sense the work of scientists may become more intelligible. There seems to be current a misapprehension that science has become so mathematical, so abstract, and so technical that any relation it had with common sense has long since been severed. To be sure, the revolution in physics in the twentieth century has come about largely because scientists have examined more thoughtfully than they did before certain common-sense ideas. This re-examination has resulted in a drastic revision of the physicists' notions about space and time. But the layman is almost certain to become hopelessly confused if he attempts to initiate his inquiry into the nature of science by starting with relativity and quantum phenomena. What is suggested here is the reverse. We start with the historic connection between common sense and science and follow these implications to the aspects of modern science (which are manifold) where the relation has hardly changed in this century. The cautious or skeptical approach and the analysis of methods of science here presented will, I trust, prove no obstacle to those who later may wish to explore by further reading any aspect of modern science.

The results of the work of physicists of this century make it difficult to write as did scientific popularizers of the nineteenth century. There is no doubt about it, somewhere about 1900 science took a *totally unexpected* turn. There had previously been several revolutionary theories and more than one epoch-making discovery in the history of science, but what occurred between 1900 and, say, 1930 was something different; it was a failure of a general prediction about what might be confidently anticipated from experimentation. This episode in itself is for me sufficient justification for treating all scientific theories and explanations as highly provisional. Indeed, for many who understand the implications of the changed attitude of physicists, the definition of science in terms of "conceptual schemes that arise from experiment and are fruitful of experiments" will seem to be the only type of definition which can pass muster at the present moment.

Professor Bridgman has recently stated that "since the turn of the century the physicist has passed through what amounts to an intellectual crisis forced by the discovery of experimental facts of a sort which he had not previously envisaged, and *which he would not even have thought possible.*" (The italics are mine.) Professor Bridgman was speaking primarily of the phenomena in the realm of high velocity for which the special theory of relativity now provides so suitable a conceptual scheme. What has happened in connection with the study of light (both visible and invisible) is, from the point of view of one who is not a physicist but has been privileged to listen to physicists for over forty years, quite as extraordinary. Briefly, the story is this: Many simple optical phenomena can be accounted for in terms of a theory (a conceptual scheme) in which light is envisaged as a wave motion; some of the same phenomena can be accounted for in terms of a theory in which a beam of light is thought of as a stream of minute particles. The corpuscular theory, as the latter conceptual scheme is called, was once a strong contender; about 1800, however, certain experimental phenomena were discovered which were difficult if not impossible to explain except in terms of the wave theory. By the middle of the nineteenth century it was generally considered that this theory of the nature of light had been "established." A somewhat cranky Harvard professor of the 1870's, however, used to tell his class that "the reason everyone now believes in the wave theory of light is that all those who once believed in the corpuscular theory are dead."

This was a good joke in Cambridge, Massachusetts, in 1912; but even then it was apparent that something more than old-fashioned adherence to bygone dogmas was involved when one mentioned the corpuscular nature of light. For it was clear by that time that many phenomena connected with absorption and emission of light could be satisfactorily accounted for only on the basis of a corpuscular theory. Here was a dilemma; but I doubt if anyone forty years ago questioned for a moment that it would be possible by suitable experiment to resolve the difficulty. Indeed, I can remember a Harvard authority on optics of that period telling an audience that light could *not* be both corpuscular and undulatory; that would be absurd. He declared that some crucial experiment would someday be devised to settle finally which of the rival views was right. But today, in 1950, I doubt if many physicists would dissent from the proposition that it seems highly probable that no experiment can ever be devised to resolve this dilemma. The conceptual scheme which accommodates all the known optical phenomena satisfactorily (and has done so for several decades) is one in which a corpuscular picture fits one set of experiments, a wave theory certain others; *no experiment can be envisaged that will answer a question which was once thought of prime significance, namely, is light really corpuscular or undulatory in nature?*

The article on "Light" in the 1929 edition of the *Encyclopaedia Britannica* starts with the following statement: "It might perhaps be expected that we should begin by saying what light 'really' is, and should then develop its characters from such a starting point; but this procedure is not possible, since light is essentially more primitive than any of the things in terms of which we might try to explain it. The nature of light is only describable by enumerating its properties and founding them on the simplest possible principles. As these principles transcend our ordinary experiences they must be cast in a purely logical, that is to say mathematical, form. . . . We shall therefore describe, largely by means of analogies, the behaviour of light, and this *is* the 'real' nature of light."

It is interesting to compare this quotation from a distinguished physicist (which I submit as evidence for the attitude adopted throughout this book) with the introduction to the same article in the 1911 edition (by an astronomer, to be sure, but reflecting the physicists' view of that time). While "light" may be defined subjectively, the author states that "the objective definition, or the 'nature of Light' is the ultima Thule of optical research." The shift in emphasis from this concern with the real "nature of light" to the readiness to equate its behavior with its reality is evident even to a reader who knows nothing about physics. Possibly those well acquainted with modern physics who chance to read these pages will feel that I have skipped rather cavalierly over a complex historical development. The contrast between the attitudes of scientists about the nature of light is intended to bring out the extreme difficulties of defining science today in terms which were commonly used fifty years ago.

To those of my scientific friends who may object to the

skeptical approach to science that runs throughout these pages, I suggest the difficulty of talking in terms of reality when we are forced to be so cautious in regard to such an apparently simple question as "what is light *really?*" Furthermore, I may point out as a matter of pure pedagogy that it is far easier for those unfamiliar with the sciences to proceed from skepticism to dogma than vice versa. Since almost all popular presentations of science today are cast in an extremely dogmatic form, there are plenty of forces at work that will restore the reader's belief in the "reality" of scientific findings if a need for such restoration seems to him desirable when he closes this volume. Finally, I note that even those who treat the findings of physical science with the least skepticism would freely admit that at every period in the early development of a new theory all but the most enthusiastic proponents have adopted an attitude of suspended judgment. Therefore, in following my exposition of the methods of experimental science by using examples from the past, the reader must be asked to become as much of a skeptic about scientific explanations as he can.

INCREASING THE ADEQUACY OF CONCEPTUAL SCHEMES

But to come back to common sense, it is hardly necessary to argue that experimentation for scientific purposes developed from experimentation for practical purposes. (We shall shortly examine the differences and similarities between an experiment in science and the trial-and-error method by which the cavemen learned how to make stone axes.) But more fundamental than this obvious relation between science and common sense is the connection between the attempts of generations of scientists to develop and improve a series of conceptual schemes connected with experimentation and the process by which an infant learns to find its way around objects and personalities. Assuming the current evolutionary theories, we may likewise relate scientific experimentation to those activities by which primitive man gradually learned to manipulate his external world and through language communicate the results of his experiments to succeeding generations.

William James long ago wrote: "The intellectual life of man consists almost wholly in his substitution of a conceptual order for the perceptual order in which his experience originally comes. . . . Every new book verbalizes some new concept, which becomes important in proportion to the use that can be made of it. Different universes of thought thus arise, with specific sorts of relation among their ingredients. The world of common sense 'things'; the world of material tasks to be done; the mathematical world of pure form; the world of ethical propositions; the worlds of logic, of music, etc., all abstracted and generalized from long-forgotten perceptual instances, from which they have as it were flowed out, return and merge themselves again in the particulars of our present and future perception. . . . Percepts and concepts interpenetrate and melt together, impregnate and fertilize each other. Neither, taken alone, knows reality in completeness. We need them both as we need both our legs to walk with."

Experimental science can be thought of as an activity which increases the adequacy of the concepts and conceptual schemes which are related to certain types of perception and which lead to certain types of activities; it is one extension of common sense. For common sense in turn may be thought of as a series of concepts and conceptual schemes which have proved highly satisfactory for the practical uses of mankind. Some of these concepts and conceptual schemes were carried over into science with only a little pruning and whittling and for a long time proved useful. As the recent revolutions in physics indicate, however, many errors can be made by failure to examine carefully just how these common-sense ideas should be defined in terms of what the experimenter plans to do. Even in connection with the simple examples in the following pages some hint as to these difficulties will be given. But in what manner the theories and experimental findings of this century bear on the broad philosophic questions of man's knowledge of the universe must be left for the reader to explore through other books and articles.

CERTAIN COMMON-SENSE ASSUMPTIONS

In all the exposition in the following chapters we shall take for granted much of the conceptual scheme of common sense. We shall take for granted the existence of other personalities with whom we can communicate, the existence of objects located in a space which is at least approximately three dimensional in the sense described by the geometry one learns at school; we shall further assume that the existence of these objects is independent of the presence of an observer. The assumption of a uniformity in nature—a belief in the reproducibility of phenomena—is basic to all science; here again common-sense notions provide the stable platform on which we build. The early experimenters of the Renaissance as well as their predecessors in the prenatal period of science in ancient times took for granted the uniformity of nature. Yet if one were to trace the intellectual history of Europe from 1300 to 1700 a great many pages would be needed to portray the changing point of view of learned men about the so-called "ob-

jective world." Therefore, at the risk of incurring the wrath
of those whose chief concern is with the metaphysical
foundations of seventeenth-century science, I may hazard
the opinion that common-sense experience—a knowledge of the practical behavior of physical objects—was a
necessary (but not sufficient) premise for experimental
science.

To be sure, Agricola in his sixteenth-century account of
mining gives as detailed a description of the gnomes and
elves who lived in the mine as he does of mining methods
and machinery. But he says that contrary to some beliefs
these spirits play no part in the workaday business of get-
ting ore to the surface; in short, he gives them no opera-
tional validity. The reproducibility of the performance of
cannon had to be assumed by military commanders long
before Galileo became interested in the trajectory of the
cannon ball. The anatomists of Padua who in the sixteenth
century did so much to forward one type of scientific inquiry knew within what limits the structure of human
bodies could be safely predicted from one cadaver to the
next. In the same way common-sense observations stretch-
ing back through the ages were the bases of agriculture.

Some historians have maintained that a metaphysical
system was the essential framework without which the
new experimental philosophy of the seventeenth century
could not have developed. Certain modern writers claim
that no atheist or agnostic could have had sufficient faith
in the uniformity of nature to undertake scientific work.
Analyzing the various components in a man's unconscious
assumptions is a difficult business. Admittedly, a belief
in a high degree of uniformity in nature is implicit in all
scientific formulations, but this faith must be at least as
powerful in those who spend their lives in carrying on the

practical arts; and this was true long before the dawn of experimental science three centuries ago.

In the following chapters we shall consider experiments which have been often repeated with the same results (within certain limits of error); we shall assume that under the same set of conditions the phenomena are in all details reproducible. Such assumptions may be regarded as an act of faith but hardly more so than similar assumptions made by the smelter of ore, the agriculturist, the navigator, the operator of pumps, the glass blower, and countless other artisans. Practical knowledge had been verified for generations before learned men became interested in the manipulation of physical objects. Indeed, I venture to use the slippery word "fact" throughout this book to designate not only a past experiment but the generalization that is involved in the prediction that under a given set of conditions the experiment can be repeated. Thus one can hardly avoid saying that it is a fact that a suction pump will draw water up to a height of about 34 feet if operated at sea level (and no higher). But in line with the cautious approach already mentioned, I shall try to avoid saying that it is a fact that "the earth in which we live is surrounded by a sea of air which by virtue of its weight exerts pressure," though personally I have no doubt about it! In an exploration such as that to which we are committed, the statement in quotation marks is best referred to as a conceptual scheme. Certainly when it was first put forward it differed in no essential respect from such a statement as "the nucleus of an atom is composed of protons and neutrons," a proposition that would today still be regarded as a hypothesis or theory (and not a fact) by many scientists.

One sometimes reads in current popular articles about

science some such declaration as the following: "Modern physics has shown that a wooden table is *really* composed of electrons, protons, and neutrons." The word "really" as used in this sentence is a word that has overtones which can be highly misleading. One is on safer ground if one says, "The concept of a table is a useful one in the common-sense world and has been used without much difficulty by all but professional philosophers (when they are philosophizing); for many practical purposes the concept of 'wood' is also useful and sufficiently defined in terms of the past history of the material. Some of the chemical transformations of wood can be usefully formulated in terms of two materials known as 'cellulose' and 'lignin,' and still more profitably accounted for in terms of the conceptual scheme of atoms and the grouping of atoms in molecules. However, there is nothing that I am aware of that one can do with wood that makes it at all profitable to bring in electrons, protons, or neutrons. (If one were talking about uranium, that would be, of course, another story!)"

In short, the question of the reality of many of the conceptual schemes of the scientist poses no more difficult problem than the question of the reality of the common-sense concept of a table or the material we call wood. For scientists as for laymen, when they are not on their guard the degree of reality is largely a function of the degree of familiarity with the concept or conceptual scheme; this in turn is a function of the fruitfulness of the idea over a considerable period of time. As to the certainty of prediction in science as in matters of common sense, it appears to be wholly a question of the degree of probability. In most of our practical behavior we are ready to bet our last dollar on such certainties as that a stone tossed in the air will fall toward the ground. One school of philosophers would say that our predictions of everyday events of which we were "certain" were instances of a knowledge with a high degree of probability. Nevertheless statements about the past, predictions about the future, generalizations about what event will follow another are all grist for the mill of the thoroughgoing skeptic. In Chapter X of this book some of the consequences of varying degrees of skepticism will be noted briefly. For an understanding of the way a modern scientist operates in his laboratory, animal faith (to use Santayana's phrase) in a world of men and objects coupled with skepticism about science will best serve the inquisitive reader. At all events, some such mixture of credulity and incredulity will be the basis of the accounts which follow of the labors of a few distinguished men of science.

ACCUMULATIVE KNOWLEDGE

Before leaving our discussion of the question "What is science?" we must consider the point of view of those who wish to use the word to cover a vast group of human activities. This is the sense in which the German word *Wissenschaft* is used. I have already indicated why I believe it to be more profitable to restrict the use of the word science. To designate the larger field I prefer the phrase "accumulative knowledge." Under the heading "accumulative knowledge" one can include all the physical and biological sciences and likewise mathematics, symbolic logic, philology, archaeology, anthropology, and large portions of historical studies as well. One can state with assurance that great advances have been made in these subjects in the last three centuries. A similar statement cannot be made about philosophy, poetry, and the fine arts. If you are inclined to doubt this and raise the question of how progress even in academic matters can be defined, I would respond by asking you to perform an imaginary operation. Bring back to life the great figures of the past who were identified with the subjects in question. Ask them to view the present scene and answer whether or not in their opinion there has been an advance. No one can doubt how Galileo, Newton, Harvey, or the pioneers in anthropology and archaeology would respond. It is far otherwise with Michelangelo, Rembrandt, Dante, Milton, or Keats. It would be otherwise with Thomas Aquinas, Spinoza, Locke, or Kant. We might argue all day whether or not a particular artist or poet or philosopher would feel the present state of art or poetry or philosophy to be an advance or a retrogression from the days when he himself was a creative spirit. There would be no unanimity among us, and more significant still, no agreement between the majority view which might now prevail and that which would have prevailed fifty years ago.

I recognize how dangerous it is to introduce the concept of progress as a method of defining an area of intellectual activity. Therefore I hasten to say that I am not implying any hierarchy in my classification. I put no halo over the words advance or progress; quite the contrary. In terms of their importance to each of us as human beings, I think the very subjects which fall outside of my definition of accumulative knowledge far outrank the others. To amplify this point would be to digress too far. I need only ask two questions: How often in our daily lives are we influenced in important decisions by the results of the scientific inquiries of modern times? How often do our actions fail to reflect the influence of the philosophy and poetry which we have consciously and unconsciously imbibed over many years? A dictator wishing to mold the thoughts and actions of a literate people might be able to afford to leave the scientists and scholars alone, but he must win over to his side or destroy the philosophers, the writers, and the artists.

PROGRESS IN KNOWLEDGE AND PRACTICE

But to return to the question of accumulative knowledge. If the boundaries of this field of human endeavor are to be measured in terms of progress or the lack of it, must we not include a vast amount of practical as well as theoretical knowledge? Undoubtedly. Indeed, to the nonacademic mind the word progress as applied to man's activities immediately evokes thoughts of synthetic drugs and cars and radios, not Newton's three laws of motion or the quantum theory or Einstein's equation. The distinction between improvement in the practical arts and advances in the sciences will be one of the recurring topics in this book. The difference between invention and scientific discovery may in a few instances seem slight, but a confusion between the history of the practical arts and the development of science is a fruitful source of misunderstanding about science. The distinction between advances in mechanical contrivances or chemical processes (such as metallurgy or soap making) and advances in science will be considered at some length in subsequent pages. In the biological sciences it is not industry but commerce and agriculture on the one hand and medicine on the other which are related to scientific progress.

The practical arts for a long time ran ahead of science; only in very recent years have scientific discoveries affected practice to a greater extent than practice has affected science. The late Professor L. J. Henderson was fond of remarking that before 1850 the steam engine did more for science than science did for the steam engine.

There can be no doubt, of course, that knowledge has been accumulated, classified, and digested to serve practical ends ever since the dawn of civilization. Relatively little is to be learned about the tactics and strategy of science by studying the history of these advances. For they do not form a part of science, though without them science would never have arisen.

Similarly, it may be argued that the progress which has been made in recent centuries in certain nations in the art of government, in the treatment of criminals, in the spread of education, in the mitigation of unequal opportunities, and in social reform in general is no part of the social sciences. Such changes bear the same relations to the science of man as do the improvements in manufacturing processes and methods of transportation to the physical sciences (though there seems to be less unanimity today as to the desirability of certain of the social changes). If this analogy be correct, those interested in the methods of the social sciences might study with profit the way the physical and biological sciences have evolved from the everyday activities of artisans and agriculturists. A study of the evolution of modern medicine from the empirical practices of former days may be of special value.

It will be noted that the definition of science here presented does not exclude the study of man. Nevertheless, I shall not venture far into a discussion of the problems and methods of the social sciences. The following pages are almost exclusively devoted to the physical and biological sciences. But in connection with the study of the past (Chapter X) I shall venture to distinguish between science and history. If the distinction be valid, then it may well have relevance to those studies which for academic purposes are often listed under the heading "social sciences." The whole area of accumulative knowledge may be divided in many ways, and it would seem to me that history, like mathematics, should occupy a position by itself. But there are bound to be uncertainties in any classification. Psychology, for example, is usually thought of as both a biological science and a social science; anthropology likewise can be considered as related to either biology or sociology. To the extent that the study of man is similar to the study of other animals, the methods of the biological sciences would seem to be clearly applicable. On the other hand, social psychology, social anthropology, and sociology represent an effort to introduce new concepts and new methods into the art of human relations and may be considered a quite separate enterprise. To what extent there is a parallelism between the procedures thus employed and those characteristic of the natural sciences is still somewhat of an open question. I shall return to it in the concluding chapter and join in a plea for further support for the scientific study of man as a social animal. We who have high hopes for progress in this undertaking believe there is to some degree an analogy with the natural sciences. If that be true, the cases presented in this book, though concerned only with physics, chemistry, and biology may be of significance to investigators who seek a better understanding of man and the societies he creates.

CHAPTER THREE

Concerning the Alleged
Scientific Method

Fᴿᴏᴍ ᴀᴛᴛᴇᴍᴘᴛs to define science we now turn to a no less controversial subject—the methods of science. Those who favor the use of the word science to embrace all the activities of the learned world are inclined to belief in the existence of *a* scientific method. Indeed, a few go further and not only claim the existence of *a* method but believe in its applicability to a wide variety of practical affairs as well. For example, a distinguished American biologist declared not long ago that "Men and women effectively trained in science *and in the scientific method,* usually ask for the evidence, almost automatically." He was referring not to scientific matters but to the vexing problems which confront us in everyday life—in factories, offices, and political gatherings.

One cannot help wondering where the author of such a categorical statement obtained his evidence. But this is perhaps making a debater's point. The significance of the statement is that it reflects a persistent belief in the correctness of the analysis of science presented by Pearson in *The Grammar of Science*. Throughout the volume Karl Pearson refers to science as the classification of facts, and in his summary of the first chapter he writes as follows: "The scientific method is marked by the following features: (a) careful and accurate classification of facts and observation of their correlation and sequence; (b) the discovery of scientific laws by aid of the creative imagination; (c) self-criticism and the final touchstone of equal validity for all normally constituted minds." With (b) and (c) one can have little quarrel since all condensed statements of this type are by necessity incomplete, but from (a) I dissent entirely. And it is the point of view expressed in this sentence that dominates Pearson's whole discussion. It seems to me, indeed, that one who had little or no direct experience with scientific investigations might be completely misled as to the nature of the methods of science by studying this famous book.

If science were as simple as this very readable account would have us believe, why did it take so long a period of fumbling before scientists were clear on some very familiar matters? Newton's famous work was complete by the close of the seventeenth century. The cultured gentlemen of France and England in the first decade of the eighteenth century talked in terms of a solar system almost identical with that taught in school today. The laws of motion and their application to mechanics were widely understood. This being the case it might be imagined that the common phenomenon of combustion would have been formulated in terms of comparable clarity once people put their minds on scientific problems. Yet it was not until the late 1770's that the role of oxygen in combustion was discovered. Another hotly debated problem, the spontaneous generation of life, was an open question as late as the 1870's. Darwin convinced himself and later the scientific world and later still the educated public of the correctness of the general idea of evolution because of his theory as to the mechanism by which evolution might have occurred. Today the basic idea of the evolutionary development of higher plants and animals stands almost without question, but Darwin's mechanism has been so greatly altered that we may say a modern theory has evolved. And we are no nearer a solution of the problem of how life originated on this planet than we were in Darwin's day.

The stumbling way in which even the ablest of the scientists in every generation have had to fight through thickets of erroneous observations, misleading generalizations, inadequate formulations, and unconscious prejudice is rarely appreciated by those who obtain their scientific knowledge from textbooks. It is largely neglected by those expounders of the alleged scientific method who are fascinated by the logical rather than the psychological aspects of experimental investigations. Science as I have defined the term represents one segment of the much larger field of accumulative knowledge. The common characteristic of all the theoretical and practical investigations which fall within this framework—a sense of progress—gives no clue as to the *activities* of those who have advanced our knowledge. To attempt to formulate in one set of logical rules the way in which mathematicians, historians, archaeologists, philologists, biologists, and physical scientists have made progress would be to ignore all the vitality in these varied undertakings. Even within the narrow field of the development of "concepts and conceptual schemes from experiment" (experimental science) it is all too easy to be fascinated by oversimplified accounts of the methods used by the pioneers.

To be sure, it is relatively easy to deride any definition of scientific activity as being oversimplified, and it is relatively hard to find a better substitute. But on one point I believe almost all modern historians of the natural sciences would agree and be in opposition to Karl Pearson. There is no such thing as *the* scientific method. If there were, surely an examination of the history of physics, chemistry, and biology would reveal it. For as I have already pointed out, few would deny that it is the progress in physics, chemistry, and experimental biology which gives everyone confidence in the procedures of the scientist. Yet, a careful examination of these subjects fails to reveal any *one* method by means of which the masters in these fields broke new ground.

THE BIRTH OF EXPERIMENTAL SCIENCE IN THE SEVENTEENTH CENTURY

As I interpret the history of science, the sudden burst of activity in the seventeenth century which contemporaries called the "new philosophy" or the "experimental philosophy" was the result of the union of three streams of thought and action. These may be designated as (1) speculative thinking (2) deductive reasoning (3) cut-and-try or empirical experimentation. The first two are well illustrated by the writings of the learned men of the Middle Ages. The professor of law and theology as well as the teacher of mathematics and logic from the eleventh to the seventeenth century was concerned with a rational ordering of general ideas and the development of logical processes. In so doing, they extended to some degree the philosophical and mathematical ideas of the ancient Greeks and laid the foundations for the science of mechanics, the first of the branches of physics to take on modern dress.

A simple illustration of deductive reasoning is to be found in recalling one's experience in school with plane geometry. A set of postulates or axioms is given; then by logical processes of deduction many conclusions follow. Similarly, less formal and rigid general ideas—speculative ideas—can be manipulated by logical procedures which, however, frequently lack the rigor of mathematical reasoning. The discussion of general speculative ideas and the more detailed handling of mathematics, it should be noted, involve processes of thought which are believed to be sufficient unto themselves. No one feels impelled to appeal to observation in building a purely rational system of ideas.

The sudden burst of interest in the seventeenth century in the new experimental philosophy was to a considerable extent the result of a new curiosity on the part of thoughtful men. Practical matters ranging from agriculture and medicine to the art of pumping, the working of metals, and the ballistics of cannon balls began to attract the attention of learned professors or inquiring men of leisure. The early history of science is full of examples where the observation of a practical art by a scientist suggested a problem. *But the solution of a scientific problem is something quite different from the advances which had hitherto been made by the empirical experimentation of the agriculturist or the workman.* The new element which was introduced was the use of deductive reasoning. This was coupled with one or more generalizations often derived from speculative ideas of an earlier time. The focus of attention was shifted from an immediate task of improving a machine or a process to a curiosity about the phenomena in question. New ideas or concepts began to be as important as new inventions. The experimentation of the skilled artisans or the ingenious contriver of machines and processes became joined to the mathematical mode of reasoning of the learned profession. But it took many generations before deductive reasoning and experimentation could be successfully combined and applied to many areas of inquiry.

SPECULATIVE IDEAS, WORKING HYPOTHESES, AND CONCEPTUAL SCHEMES

Science we defined in the last chapter as "an interconnected series of concepts and conceptual schemes that have developed as a result of experimentation and observation and are fruitful of further experimentation and observations." A conceptual scheme when first formulated may be considered *a working hypothesis on a grand scale*. From it one can deduce, however, *many* consequences, each of which can be the basis of chains of reasoning yielding deductions that can be tested by experiment. *If these tests confirm the deductions in a number of instances, evidence accumulates tending to confirm the working hypothesis on a grand scale, which soon becomes accepted as a new conceptual scheme.* Its subsequent life may be short or long, for from it new deductions are constantly being made which can be verified or not by careful experimentation.

In planning the experiments to test the deductions it became necessary, as science advanced, to make more precise and accurate many vague common-sense ideas, notably those connected with measurement. Old ideas were clarified or new ones introduced. These are the new concepts which are often quite as important as the broad conceptual schemes. It is often much more difficult than at first sight appears to get a clear-cut yes or no answer to a simple experimental question. And the broader hypotheses must remain only speculative ideas until one can relate them to experiment.

An understanding of the relationship between broad speculative ideas and a wide conceptual scheme is of the utmost importance to an understanding of science. A good example is furnished by the history of the atomic theory. The notion that there were fundamental units—ultimate particles—of which matter was composed goes back to ancient times. But expressed merely in general terms this is a speculative idea and can hardly be considered an integral part of the fabric of science until it becomes the basis of a working hypothesis on a grand scale from which deductions capable of experimental test can be made. This particular speculative idea or working hypothesis on a grand scale became a new conceptual scheme only after Dalton had shown, about 1800, how fruitful it was in connection with the quantitative chemical experimentation that had been initiated by the chemical revolution. Here is an instance where we can see in some detail the origins of a working hypothesis, while in other instances we are uncertain how the idea came to the proponent's mind.

The great working hypotheses in the past have often originated in the minds of the pioneers as a result of mental processes which can best be described by such words as "inspired guess," "intuitive hunch," or "brilliant flash of imagination." Rarely if ever do they seem to have been the product of a careful examination of all the facts and a logical analysis of various ways of formulating a new principle. Pearson and other nineteenth-century writers about the methods of science largely overlooked this phenomenon. They were so impressed by the classification of facts and the drawing of generalizations from facts that they tended to regard this activity as all there was to science. Nowadays the pendulum has swung to the other extreme and some writers seem to concentrate attention on the development of new ideas and their manipulation, that is on theoretical science. Both points of view minimize the significance of the experiment. To my mind this distorts the history of science and, what is worse, confuses the layman who is interested in the scientific activity which is going on all about him. For these reasons and because of the author's own predilection, the present discussion of science and common sense emphasizes and re-emphasizes the interrelation of experiments and theory.

EXPERIMENTATION

The three elements in modern science already mentioned are: (1) speculative general ideas, (2) deductive reasoning, and (3) experimentation. We have discussed in a general way the manner in which new working hypotheses on a grand scale arise and how from them one may deduce certain consequences that can be tested by experiment. It has been implied that the art of experimentation long antedates the rise of science in the seventeenth century. If so, this is one way in which science and common sense are connected. A rather detailed analysis of experimentation in everyday life may serve a useful purpose at this point. For, as I hope to show, there is a continuous gradation between the simplest rational acts of an individual and the most refined scientific experiment. Not that the two extremes of this spectrum are identical. Quite the contrary: to understand science one must have an appreciation of just how common-sense trials differ from experiments in science.

Since the reader may well have been exposed at some

time in his or her life to various statements about the alleged scientific method, it may be permissible to set up a few straw men and knock them down. I have read statements about the scientific method which describe fairly accurately the activity of an experimental scientist on many occasions (but not all). They run about as follows: (1) a problem is recognized and an objective formulated; (2) all the relevant information is collected (many a hidden pitfall lies in the word "relevant"!); (3) a working hypothesis is formulated; (4) deductions from the hypothesis are drawn; (5) the deductions are tested by actual trial; (6) depending on the outcome, the working hypothesis is accepted, modified, or discarded.

If this were all there was to science, one might say, in the words of a contemporary believer in *the* scientific method, that science as a method "consists of asking clear, answerable questions in order to direct one's observations which are made in a calm, unprejudiced manner, reported as accurately as possible and in such a way as to answer the questions that were asked to begin with; any assumptions that were held before the observations are now revised in the light of what has happened." But if one examines his own behavior whenever faced with a practical emergency (such as the failure of his car to start) he will recognize in the preceding quotation a description of what he himself has often done. Indeed, if one attempts to present the alleged scientific method in any such way to a group of discerning young people they may well come back with the statement that they have been scientists all their lives! The layman confronted with some such description of science is in a similar situation to that of the famous character in Molière's comedy who had been speaking prose all his life without knowing it.

TESTING DEDUCTIONS BY EXPERIMENT

Is there then no difference between science and common sense as far as method is concerned? Let us look at this question by considering in some detail first an everyday example of experimentation and then a scientific investigation. The type of activity by which the practical arts have developed over the ages is in essence a trial and error procedure. The same sort of activity is familiar in everyday life; we may call it experimentation. Let us take a very restricted and perhaps trivial example. Faced with a locked door and a bunch of keys lying on the floor, a curious man may wish to experiment with the purpose of opening the door in question. He tries first one key and then the other, each time essentially saying to himself, "If I place this key in the lock and turn it, then the result will either confirm or negate my hypothesis that this key fits the lock." This "if . . . then" type of statement is a recurring pattern in rational activity in everyday life. The hypothesis that is involved in such a specific trial is limited to the case at hand, the particular key in question. We may call it therefore a *limited working hypothesis.*

Let me turn now from an example of commonplace experimentation to a consideration of a scientific experiment. Let us examine the role of the limited working hypothesis in the testing of some scientific idea in the laboratory. For if you brought all the writers on scientific method with the most varied views together, I imagine that every one of them would agree that the testing of a deduction from a broad working hypothesis (some would say a theory) was at least a part of science. In the next chapter we are going to consider in some detail several instances of such procedures. Let us anticipate the story about atmospheric pressure only to the extent of fixing our minds on one actual experiment. It makes little difference which one we choose, for we wish to center attention on the last step, the actual experimental manipulation.

We will imagine that someone has related the broad hypothesis that we live in a sea of air that exerts pressure to a particular experiment with a particular piece of apparatus. Just before turning a certain stopcock, which we may assume to be the final step in the experiment, the investigator may formulate his ideas in some such statement as, "If all my reasoning and plans are right, when I turn the stopcock, such and such will happen." He turns the stopcock and makes the observation; he can then say he has confirmed or failed to confirm his hypothesis. But strictly speaking, let it be noted, it is only a highly limited hypothesis that has been tested by turning the stopcock and making the observation; this hypothesis may be thrown into some such form as "if I turn the stopcock, then such and such will happen." The confirmation or not of this extremely limited hypothesis is regarded as an experimental fact if repetition yields the same result. The outcome of the experiment is usually connected with the main question by a highly complex process of thought and action which brings in many other concepts and conceptual schemes. The examination of such processes is involved in a study of the examples of "science in the making" in the following chapters. The point to be emphasized here is the existence of a complicated chain of reasoning connecting the consequence deduced from a broad hypothesis and the actual experimental manipulation; furthermore, we shall repeatedly see how many assumptions, some conscious and some unconscious, are almost always involved in this chain of reasoning.

Now perhaps I may be permitted to jump from the scientist to the everyday experiments of a householder in his garage or a housewife in her kitchen or an amateur with his radio set. If a car won't start, we certainly have a problem; we think over the various possibilities based on our knowledge of automobiles in general and the particular car in question; we construct at least one working hypothesis (the gas tank is empty!); we proceed to carry out a trial or test (an experiment) which should prove the correctness of this particular working hypothesis; if we are right then we believe we have found the trouble and proceed accordingly. (But how often have we been misled; perhaps there is more than one trouble; perhaps the tank is empty and the battery run down too!) Let us assume the simple hypothesis has led to a trial which consists of turning a particular switch or making a certain connection of wires after various other manipulations have been performed. Then one says to oneself, "Now at last if I turn the switch (or make the connection) the engine will turn over." The test is made and what is confirmed (or not) is a very limited working hypothesis hardly to be distinguished from the extremely limited working hypothesis of the scientists we have just been considering. Here science and common sense certainly seem to have come together. But notice carefully, they are joined only in the statement of the final operation. As we trace back the line of argument the differences become apparent. These are differences as to aims, as to auxiliary hypotheses and as to assumptions.

AIMS AND ASSUMPTIONS IN SCIENTIFIC EXPERIMENTATION

First, as regards aims, you want the car to start (or the radio set to work, to mention another example); you wish to reach a practical objective. The scientific experimenter on the other hand wants to test a deduction from a conceptual scheme (a theory), a very different matter. But we cannot leave the distinction between science and common sense resting on that point, important as it is. The conceptual scheme is not only being tested by the experimenter but it has given rise to the experiment. And that takes us back to our definition of science and our emphasis on the significance of the fruitfulness of a new conceptual scheme. The artisans who improved the practical arts over the centuries proceeded much as you or I do today when we are confronted with a practical problem. The aim of the artisan or the agriculturist was practical, the motivation was practical, though the objective was more general than just starting a given automobile. The workmen of the Middle Ages experimented and sometimes their results left a permanent residue because their contemporaries incorporated a new procedure into an evolving art. But the artisan rarely, if ever, bothered about the testing of the consequences of any general ideas. General ideas, logical thought, were the province of learned men. Drawing deductions from conceptual schemes was the type of activity known to the mathematicians and philosophers of the Middle Ages, not to the workmen. And with rare exceptions those who understood these recondite matters paid no attention to the workmen. In the next two chapters we shall consider examples of how the two activities—those of the logicians and the artisans—came together in the sixteenth and seventeenth centuries.

There is another important difference between the artisan and the scientist. The typical procedure of the artisan is very much like that of the housewife in the kitchen.

Not only are the trials of new procedures for very practical and immediate ends, but the relevant information is largely unconnected with general ideas or theories. Until as late as the nineteenth century the practical man paid very little attention to the growing body of science. By and large the practical arts and science went their own ways during the seventeenth and eighteenth centuries. We may say that experimentation in the practical arts or in the kitchen is almost wholly empirical, meaning thereby to indicate the absence of a theoretical component. However, since the transition from common sense to science is gradual and continuous, one may well question if there is ever a total absence of a theoretical background. It can be argued that the concepts and conceptual schemes we take for granted in everyday life and share in common with our ancestors are not different in principle from the "well-established" ideas of science. This I believe to be true in the sense that infrared light is not different in principle from X rays; both forms of radiant energy are parts of a spectrum, but the two forms of light are certainly not interchangeable for most practical purposes. Quite the contrary. So too, common-sense ideas are distinct in many respects from the more abstract part of the scientific fabric. During the last two hundred years more and more of the material of science has become incorporated into our common-sense assumptions. But every age and cultural group has its own way of looking at the world. The pictures of the total universe which can be collected by anthropologists and students of cultural history, while having many assumptions in common, likewise show great divergences. Therefore, if the modern man in his garage "takes for granted" many things his grandfather would have believed impossible, this in no way invalidates the differentiation between common-sense ideas and scientific theories. (Though there is a wide, fuzzy, intermediary zone.)

THE DEGREE OF EMPIRICISM IN A SCIENCE OR A PRACTICAL ART

In analyzing the present relation of science to technology and medicine I have found it useful to use the term "degree of empiricism" to indicate the extent to which our knowledge can be expressed in terms of broad conceptual schemes. The same phrase is likewise useful, I am inclined to think, in connection with the history of both the sciences and the practical arts in the last three hundred years. The importance of the notion which lies back of the term, however, is that it may be of real help to the layman who is confused about the relation of "pure" and "applied" science. In the last one hundred years science and technology have become so intertwined that even the practitioners in the field may be uncertain when they attempt to analyze the role of scientific theories. Yet anyone familiar with the physical sciences and modern industry would at once grant that there were wide differences in the extent to which scientists in different industries can apply scientific knowledge to the work at hand.

To illustrate what to me is a highly important point, let me contrast the business of making optical instruments with that of manufacturing rubber tires. The design of lenses and mirrors for telescopes, microscopes, and cameras is based on a theory of light which was developed 150 years ago and which can be expressed in simple mathematical terms. With the aid of this theory and a few measurements of the properties of the glasses used, it is possible to calculate with great accuracy the performance of optical equipment.

Since theoretical knowledge is so complete in the field of optics, we may say that the degree of empiricism is low in this branch of physics. Because theory is so effective in the optical industry, the degree of empiricism is low. The manufacture of rubber tires is a totally different story. There is nothing comparable to the theory of light to provide a mathematical basis for calculation of what ingredients should be mixed with the rubber. The chemical change which is basic to the whole process is known as vulcanization but no one is very clear even today how to formulate it in theoretical terms. The action of the sulfur which was long thought to be an essential ingredient and of certain other chemicals known as "accelerators" is but little understood. The whole process has been built up by trial and error, by a procedure in which a vast number of experiments finally yielded knowledge in every way comparable to the knowledge of a first-rate chef. In this industry the degree of empiricism is high, and this reflects in turn the small extent to which the chemistry of rubber has been formulated in wide theoretical terms.

As in all comparative statements we must have some fixed points for standards. Therefore, without entering further into the philosophic analysis of common-sense knowledge (including the knowledge of an "art" such as cooking, or glass blowing, or metal making in the Middle Ages), we may take as an example of essentially empirical procedures those of the artisan before the advent of modern science and of the cook in a modern kitchen. Here one may arbitrarily say the degree of empiricism is practically 100 per cent. To find a case where the degree is so low that we can take it as zero on our scale, I call your attention to the work of the surveyor. The theoretical framework is here largely one branch of mathematics—geometry—and, except to a very slight extent in the building of the instruments and the efficiency of their handling, empirical procedures are conspicuous by their absence. Therefore if the reader of this book whose acquaintance with science or technology is slight will envisage from time to time the surveyor with his transit and his measuring instruments on the one hand and the chef in the Grand Hotel on the other, he will have in mind a range of activities from zero to 100 as regards the degree of empiricism involved.

We shall return from time to time to the relation of scientific knowledge to the practical activities of the artisan, the agriculturist, and the medical man. We shall see that for an amazingly long time advances in science and progress in the practical arts ran parallel with few interconnecting ties. Thus if we take the birth of modern science as somewhere around 1600 (neglecting the long prenatal period which goes back to antiquity), one can say that it was two hundred years or more before the practical arts benefited much from science. Indeed, it would be my contention that it was not until the electrical and dyestuff industries were well started, about 1870, that science became of real significance to industry.

Let me conclude this discussion by pointing out that the degree of empiricism in a practical field today largely depends on the extent to which the corresponding scientific area can be formulated in terms of broad conceptual schemes. Therefore one may consider science as an attempt either to lower the degree of empiricism or to extend the range of theory. When scientific work is undertaken without regard for any practical application of the knowledge, it is convenient to speak of the activity as part of "pure" science. But there are certain overtones in the adjective which are unpleasant and seem to imply a hierarchy of values as between the scientists interested in theories and those interested in the practical arts. Therefore the phrase "basic science" is frequently employed. Almost all significant work of scientists today, I believe, comes under the heading of attempts to reduce the degree of empiricism; the distinction between one group and another is in the motivation. Those who are interested in the fabric of science as such are ready to follow any lead that gives promise of being fruitful in terms of extending theoretical knowledge. Others are primarily concerned with one of the ancient practical arts in modern dress; if it is some branch of industry, say metallurgy, they will be just as interested in widening the theoretical knowledge *in this field* as their colleagues in a university; they will be endeavoring to lower the degree of empiricism too, but in a limited area for a practical objective. The medical scientist is like the metallurgist except that his goal is not better metals but healthier people; both are working in applied science.

In the 1950's, therefore, we find a complex state of affairs. About three centuries ago the trial-and-error experimentation of the artisan was wedded to the deductive method of reasoning of the mathematician; the progeny of this union have returned after many generations to assist the "sooty empiric" in his labors. In so doing the applied scientist finds himself face to face with one of his distant ancestors, so to speak. For as he works in an industrial laboratory he will often find himself called on to carry out experiments for a practical purpose on nearly as empirical a basis as the artisan of the distant past. Particularly in those practical arts where the degree of empiricism is still high, men with the most advanced scientific training and using the latest equipment will often have to resort to wholly empirical procedures. On the one hand they will labor to reduce the degree of empiricism as best they can; on the other they must improve the art by using the knowledge and methods then at hand. In short, advances in science and progress in the practical arts today go hand in hand.

In this chapter devoted to the alleged scientific method I have concentrated attention on the advance of scientific knowledge. In so doing I have had occasion to examine progress in the practical arts and particularly the trial-and-error methods of the artisan. The picture thus drawn of the advances of science and technology in the last 150 years, while illuminating, I believe, as regards science, is lacking in substance as regards technology. For I have failed to convey to the reader, I am sure, any feeling for the vast amount of labor of those who applied the science of the day to many practical undertakings. In short, I have said little or nothing about engineering. This is a lack which subsequent chapters will to some degree remedy, but only a history of the separate fields of engineering—civil, mechanical, electrical, aeronautical, and chemical—could really do justice to the modern history of the application of the physical sciences to industry. And without this history, terminology and definition have no solidity. The first engineers were military; then came the application of surveying and map making to nonmilitary purposes—civil engineering became a profession. Until well into the nineteenth century the civil engineer not only was the man who made surveys, built bridges, canals, and roads but also was concerned with machinery as well. Watt, of steam engine fame, was considered by his contemporaries as an engineer but a *civil* engineer.

The improvement of steam engines and machines of all

sorts from 1700 to the mid-nineteenth century was the work of men who called themselves inventors or engineers. The construction of new equipment and its installation were done by men of business who might also be inventors and often regarded themselves as engineers. Mechanics was by then a well-developed science with a medium degree of empiricism; therefore all these practical men were operating, much as the artisans of the Middle Ages, by trial and error and yet often with the possibility of applying theoretical principles and using mathematical calculations. Eventually the importance of this type of work led to the recognition of mechanical engineering as a special field, and electrical engineering became a designated area of the application of science to industry at about the same time (mid-nineteenth century). Today engineers by and large are concerned with designing and building machines and equipment for all sorts of practical purposes throughout industry; without them our industrialized civilization would cease. Some are more involved than others in new departures, in what is called "development work." Here they join forces with the applied scientists; or we might better say, men trained as engineers often advance applied science (reducing the degree of empiricism); conversely, men trained as scientists often operate as engineers.

My concern to give the reader some over-all understanding of the modern scene has outrun my desire to convey a better understanding of experimental science. The extent to which engineers and scientists overlap in their activities and the degree to which they can cooperate are difficult to describe to one who is unfamiliar with the procedures of the experimental investigator. Therefore I must postpone further consideration of some immediate problems of our day, such as the organization of science and engineering in government and industry, until some case histories have been examined. We must concentrate attention on the new experimental philosophy that came to life in the seventeenth century before we consider how it is interwoven with practical affairs in the times in which we live.

CHAPTER FOUR

The Development of the Concept of Atmospheric Pressure

FOR AGES people have been aware of the fact that to drain a liquid out of a barrel there should be an opening near the bottom and one at the top; out of the lower orifice comes the wine, into the upper goes the air. Similarly, everyone knows that you can suck up liquid in a tube, close the top opening with your finger, and the liquid will not run out until the finger is removed; this is the principle of the pipette (Fig. 1). Similar observations have been discussed since before the time of Aristotle. The standard explanation before the seventeenth century was couched in terms very similar to those which would be used by most of us today. "You have to have an opening in the top of the barrel to let the air in, otherwise the liquid won't run out." Of course, if pressed, the modern man (or woman) would have to admit to loose phraseology. He or she would probably say that the air coming in at the top was a consequence of the liquid running out. And sooner or later skillful and sympathetic cross-examination would elicit some information about the role of atmospheric pressure. Those who remembered their high-school physics might reply somewhat as follows: "The liquid in the pipette and in the barrel is prevented from running out by the pressure of the atmosphere; the purpose of taking your finger off the top of the pipette or opening a hole in the top of the barrel is to let

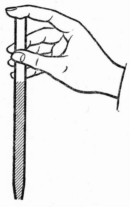

FIG. 1. A pipette. If liquid is sucked up into a small tube and a finger pressed over the top end of the tube, the liquid will not run out.

the atmospheric pressure act on the top surface of the liquid. When that is done, the atmospheric pressure is essentially the same on the top and bottom of the liquid, which then runs out for the same reason a stone falls to earth."

The learned men of the Middle Ages, however, would have followed up the common-sense appraisal of the situation in very different terms. They would have taken quite literally the statement that the hole in the top of the barrel or the opening in the top of the pipette was necessary for the air to enter. For they would have started with the assumption that the universe was full. Therefore, they maintained, there could be a movement of the liquid out of the barrel only if some place were provided for the air that had to move to make room for the liquid; the place for this air was provided by the opening in which the air could come in.

The explanation in terms of a "full universe" satisfied generations of inquirers; it was part and parcel of the view of the world which comes from the writings of Aristotle as these were interpreted by the scholars of the Middle Ages. To do justice to this world view would require many chapters. One may caricature it by picking up one of the phrases used by the Aristotelians, namely that it was a universal principle that "nature abhors a vacuum." This principle was invoked to explain why the wine would not flow out of a single opening in a barrel; it was argued that if it did and nothing entered the barrel, a vacuum would result, and this was impossible. Which is, perhaps, only a picturesque way of stating that there must be an opening to let in the air, as many a person would say in an unreflective moment in the mid-twentieth century. The common-sense view of the world even today is more Aristotelian than we sometimes like to think.

The same principle that nature abhors a vacuum could be invoked to explain why one could suck up liquid into a tube or why a lift or suction pump would raise water. Consider the action of a modern version of this very ancient device (Fig. 2), the old-fashioned pump, not long ago a feature of every kitchen. The operation of the handle raises the plunger and if the plunger is tight the water rises, is "sucked up" we often say. Why? Because otherwise there would be a vacuum, an Aristotelian would reply, and a vacuum is impossible. Again, some such explanation seems to have satisfied philosophers for generations. The first indication of a difficulty appears in the writings of Galileo. In his *Dialogues concerning Two New Sciences,* published

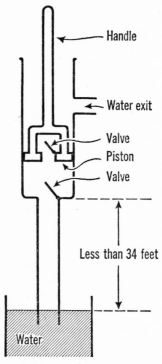

Fig. 2. Cross-sectional diagram of a simple lift or suction pump. As the piston moves upward, water rises in the pipe.

in 1638, he refers more or less in passing to the fact that a suction pump will not raise water more than a certain height. His explanation of this phenomenon need not detain us, for it was based on a poor analogy between the breaking of a column of water and the breaking of a long wire. But it is worth noting that this famous Florentine scientist missed the opportunity to make still another great contribution to the advance of science. Those who are inclined to think that the mere recognition of a problem by a great scientist will automatically produce the answer should ponder on this episode in the history of science.

A suction pump cannot raise water more than 34 feet. Galileo in discussing this phenomenon implies that it was first called to his attention by a workman. Suction pumps were no recent invention in Galileo's time. On the contrary, their history can be traced back for centuries; furthermore, practical men must have been long aware of the limitations on their use, for illustrations in Agricola's famous treatise on mining show suction pumps in tandem (Fig. 3). It is strange that no discussion of the limit in the ability of a pump to raise water has been found earlier than that in Galileo's treatment. Perhaps any who thought about the subject attributed the failure to raise water more than a certain height to mechanical imperfections; indeed, the system of plungers and valves was very crude. But I am more inclined to think that the silence is merely a striking evidence of the vast gulf that separated artisans and learned men for centuries. There were those who operated mines, smelted ores, and operated pumps; they were improving the practical arts by endless cut-and-try experimentation. Then there were the professors and the learned men at princely courts who developed mathematics, deductive reasoning, and the embryo science of mechanics. Experimental science began when these two streams of human activity converged.

What Galileo missed, his pupil Torricelli found. In 1644, six years after the publication of Galileo's fruitless interpretation and two years after the master's death, Torricelli

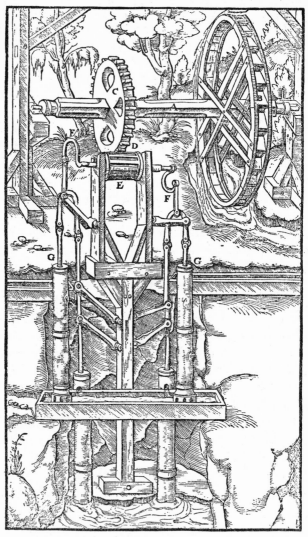

Fig. 3. An illustration from Agricola's sixteenth-century book on mining, showing the use of pumps for removing water from mines.

put down in writing some general but precise ideas about the atmosphere and atmospheric pressure. A broad hypothesis, we may call it, or a new conceptual scheme in the making. However you wish to characterize Torricelli's views, set forth in an exchange of letters with Cardinal Ricci, they represent a clean break with the Aristotelian notion of nature's abhorrence of a vacuum. In some way and at some date not recorded he saw that the limit of about 34 feet beyond which water would not rise in a suction pump might be a measure of the pressure of the atmosphere. He argued that if the earth were surrounded by a "sea of air" and if air had weight, there would be an air pressure on all objects submerged in this sea of air exactly as there is water pressure below the surface of the ocean.

Then follows a deduction from this hypothesis on a grand scale and the experimental confirmation of this deduction. If the column of water 34 feet high is sustained by the pressure of the atmosphere, the column of a liquid nearly 14 times as heavy, namely liquid mercury, should be held up only $3\frac{4}{14}$ or $2\frac{3}{7}$ feet. This is something that can be tested and it was. Again we are uncertain as to the time and place, but somewhere around 1640 and probably in Florence Torricelli performed the classic experiment with which his name will be forever connected.

If the reader has never seen the Torricellian experiment performed, he ought to visit some high-school laboratory and persuade the physics or chemistry teacher to repair his education in this respect. This is one of the few revolutionary experiments which can be carried out with the simplest of equipment and understood with the minimum of sophistication in science or mathematics (Fig. 4). Take a glass tube about a finger's width in diameter, 3 feet long, and sealed off at one end; fill it with liquid mercury (quick-

silver); then placing the thumb or forefinger over the open
end (carefully excluding any bubbles of air) invert the
tube and submerge the end closed by your finger in an open
vessel containing more mercury. Remove your finger and lo

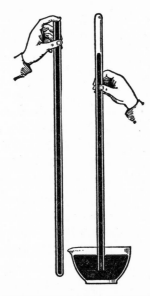

THE TORRICELLIAN EXPERIMENT

Fig. 4. The tube is completely filled with mercury, the open end
is sealed with a finger, and the tube is inverted and placed with the
"open" end in a dish of mercury. When the finger is removed, the
mercury drops until the column is about 30 inches high.

and behold, the mercury which had hitherto filled the tube
now drops, leaving an empty space in the upper end of the
tube. And the space is really empty, for you have just
produced a vacuum as Torricelli did on the day when he
first performed this experiment. Furthermore, you will have
constructed a barometer. If you live close to sea level (and

have carefully avoided air bubbles in filling the tube with the mercury) the height of the column will be approximately 30 inches. If you live at higher altitudes it will be less. Observed over a period of days, the level will fall and rise, as Torricelli soon discovered. This variation in atmospheric pressure is now a matter of daily news but it was more than a century before the relation between barometric changes and weather began to be noted.

At one stroke Galileo's disciple had invented a new instrument, tested the validity of one deduction from a working hypothesis on a grand scale, and produced a vacuum, which the Aristotelians maintained was impossible. The significance of the experiment for those studying the methods of science lies, however, in the fact that it constitutes a simple verification of *one* consequence of a very broad hypothesis or conceptual scheme. Historically we cannot be certain that the broad hypothesis preceded the experiment, for there is no record of how Torricelli came either to his idea or his experiment. Considering the recorded discussions of the phenomenon of the water pump by his great master, however, it seems extremely probable that the idea of atmospheric pressure preceded the Torricellian experiment.

One of the most baffling features of the advance of science is the unpredictable way in which revolutionary ideas have developed. Few if any pioneers have arrived at their important discoveries by a systematic process of logical thought. Rather, brilliant flashes of imagination or "hunches" have guided their steps—often at first fumbling steps. In one historic instance we can follow in some detail how all this occurred; those who persevere to Chapter VII (and through it) will be able to follow the evolution of Lavoisier's new concepts about combustion.

If we may assume the usual interpretation of Torricelli's mental processes to be correct, then one deduction from his new broad hypothesis was confirmed when he constructed the first barometer. A second important deduction was soon drawn by a French mathematician and tested by experiment. Here the pioneer was that extraordinary figure in the history of science and theology, Blaise Pascal. Hearing of the Torricellian experiment through the Parisian letter writer, Father Mersenne, he at once repeated the experiment in Rouen. He also set up a water barometer by arranging a series of tubes of water so that above a column of water 34 feet high there was a vacuum. (This, however, could be hardly considered an independent check of the Torricellian hypothesis.) The new consequence to which Pascal directed attention was this: if we live in a sea of air that exerts pressure, the situation is analogous to that which exists at the bottom of the ocean. Pascal and his contemporaries understood the phenomena connected with water pressure. The laws of hydrostatics had been formulated in the preceding century and were beautifully expounded in a volume by Pascal himself. The pressure below the surface of the water in a reservoir, lake, or ocean depends on the depth. A submarine creature rising from the ocean floor toward the surface will be subjected to a regularly *diminishing* pressure. If we live in a sea of air, the same phenomenon should be observed in the atmosphere, Pascal argued. Torricelli had provided the necessary instrument to measure the pressure—his inverted tube with a column of mercury, the barometer.

Pascal, reasoning in some such fashion, arranged for his brother-in-law Perier to carry out a series of Torricellian experiments at different heights on a mountain in central France. He considered this test a crucial one. In a letter in

1647 he said that the experiment with the inverted tube of mercury gave one reason to believe that "it is not abhorrence of the vacuum that causes mercury to stand suspended, but really the pressure of the air which balances the weight of the mercury column." Nevertheless, he went on to say, the ancient principle of the abhorrence of a vacuum can also be invoked to explain the phenomenon. But he maintained that if on carrying out the Torricellian experiment at the top and at the foot of a mountain it should turn out that the height of the column were less in the first instance than in the second, then "it follows of necessity that the weight and pressure of the air are the sole cause of this suspension of the mercury, and not the abhorrence of a vacuum, for . . . one cannot well say that nature abhors a vacuum more at the foot of the mountain than at its summit."

Pascal's brother-in-law proved to be most obliging and in September, 1648, carried out his mission. The results were as expected. The height of the column in the Torricellian tube at the top of the mountain chosen (the Puy-de-Dôme) was less than at the bottom by about three inches, and part way down the mountain the length of the column was a little more than at the top but definitely less than at the bottom. Perier further reported that the repetition of the experiment on the top gave the same result at five different points on the summit, some sheltered, some in the open, and one carried out when a cloud carrying rain drifted over the summit. An observer stationed at the bottom had been meanwhile watching one tube during the entire time; he had found that the level had remained unchanged (the atmospheric pressure during this period did not alter).

A second important deduction from the new hypothesis about a sea of air and atmospheric pressure had thus been

verified. In Pascal's mind, at least, the results were conclusive. Indeed, we might be tempted to say that after the
Puy-de-Dôme experiment Torricelli's new idea had become a new conceptual scheme! Yet with such a conclusion
as well as with Pascal's confidence, one may quarrel. The
history of science since his time has shown the dangers of
regarding the experimental verification of any *one* deduction as conclusive evidence as to the validity of a hypothesis.

It will be worth while to examine in some detail the experiment planned by Pascal and performed by Perier. First
of all, it is worth noting that the new conceptual scheme
could not be directly tested by experiment, though one can
easily fall into the habit of speaking as though this were
possible. In general, few if any hypotheses on a grand scale
or conceptual schemes can be directly tested.

A whole chain of reasoning connects a conceptual
scheme with the experimental test. This may seem a trivial
point, but it is not. More than one false step has resulted at
this point; hidden assumptions as prickly as thorns abound;
these may puncture a line of reasoning. What was observed
in an experiment may not be actually related to the conceptual scheme in the way the experimenter had believed.
To this point we shall return in other cases and show how
experimental errors can lead an experimenter astray. If
we were to attempt to expound the difficult subject of
twentieth-century physics, we should encounter the issue
in a somewhat different form. We should then find that
what appeared to be obvious common-sense assumptions
involved in the line of reasoning connecting experiments
with conceptual schemes had to be revised when either
very high speeds or very small particles were involved.

Because of their significance to an understanding of the logic of experimentation, let us examine Perier's procedures with some care. Pascal's deduction from the new conceptual scheme may be stated somewhat as follows: "If the earth is surrounded by a sea of air and if air has weight, the pressure of the air will be less on the top of a mountain than at the foot." To translate this deduction into a specific experiment requires a long line of argument. If the height of the Torricellian column is a measure of the atmospheric pressure, then on performing the experiment on the top of the Puy-de-Dôme and at the foot, the observed height will be less in the first instance than in the second, *provided* no other factors have influenced the pressure in the meantime or the height of the mercury column.

The "if" and the "provided" in this argument are of extreme importance. To guard against one obvious source of error, namely variations in the atmospheric pressure, Perier, as we have seen, had someone keep watch on a tube set up at the foot of the mountain during the entire time. To assure himself that in each case the height of the column was a true measure of the pressure, he says he carefully excluded all bubbles of air (for even a small one will on ascending into the vacuum cause the mercury to fall perceptibly). He does not state how he assured himself that he had been successful in his manipulations, and the accuracy with which his repeated observations agree with each other cannot but cause a skeptic to raise his eyebrows. Indeed, I am inclined to think he allowed his enthusiasm to affect the validity of his record. But there is no need to go into this interesting historical point; suffice it to say that standards of experimentation and recording had not been established by custom in 1648.

Perier performed certain manipulations at certain spots according to a careful set of rules. What he actually observed each time was a distance between two mercury levels; he probably used a wooden measuring stick graduated in inches and in lines (a twelfth of an inch). At the time of performing the experiment Perier was reasoning very much as the artisan or the housewife who makes a test of a new approach to a practical problem, for in effect he said, "If I perform Torricelli's experiment on this spot and there have been no errors in my manipulation and no unknown factors affecting the height of the mercury (a hypothesis limited to the case at hand), I shall observe a height of a mercury column less than is now being observed at the foot of the mountain." For all Pascal or Perier knew, a mercury column might be appreciably shorter at the top of a mountain than below for a variety of reasons: the relative densities of mercury and air might change (the ideas about gravity were then in process of formulation); the measuring stick might alter in length on being moved upward several thousand feet. Perier himself recognized that an open space and an enclosed building might make a difference, also a passing cloud. By repeating the experiment in and out of a chapel, when the air was clear and rainy, he tested the effect of certain *variables,* but the results he says were always the same.

The fact that so many variables exist in connection with testing a deduction by means of an actual experiment is the significant point. In this case, subsequent investigation has so far failed to reveal any variables that invalidate Perier's *qualitative* check on Pascal's deduction that the mercury column in a Torricellian tube is shorter on the top of a mountain than at the bottom.

YOUNG MEN AND AMATEURS: A DIGRESSION

At this point I propose to interrupt the story of seventeenth-century pneumatics to examine briefly an example of a recurring phenomenon in the history of science. I refer to the new waves of interest followed by important work that seem to sweep over the learned world from time to time. A new idea or new discoveries or new instruments open up a new area of inquiry: everyone rushes in and science advances in that field with astonishing speed, then progress lags and there may be a long period of marking time. The phenomenon to no small degree is connected, I believe, with the proverbial desire of young men to differ with their elders and to seek fields for new adventure. More than once the elderly scientists have said in essence, "Ah, that is a young man's game." The study of atmospheric pressure in the mid-seventeenth century was exactly that, for Torricelli was thirty-six when he wrote his famous letter to Cardinal Ricci; Pascal was twenty-four when he outlined the Puy-de-Dôme experiments; Boyle, whose work we are about to consider, was thirty-two when he started his line of investigation in this field.

These men, it must be remembered, were strictly amateurs; not for many years to come was experimental science to find a home in the universities; research laboratories and institutes were a full two centuries away. Galileo had been a professor in Padua, to be sure, but he represents almost the end of the scientific contributions which came from that famous center of the new learning. Boyle lived in Oxford at the time we are considering, but that university was for a brief period the exception which proved the rule. For this ancient seat of learning was then staffed by a group of young men intruded by Cromwell's army. These anti-Royalists represented varying shades of Puritanism, but all were interested in Bacon's philosophy and the experimental approach to problems. After the restoration of Charles II drastic changes in the universities were once again in order. The members of the scientific group either were replaced by the former dons who had been loyal to the king or left of their own accord. They soon entered the re-established church, and held positions of importance, but they did not return to Oxford which ceased to be a scientific center. The Royal Society, chartered by this same group a few years later, had its home in London.

In Florence the patronage of the dukes enabled another band of young amateurs—Torricelli's friends—to continue to work together after Torricelli's premature death in 1647. This Accademia del Cimento prospered from 1657 to 1667 in a completely Catholic country, let it be noted, after Galileo's famous trial and condemnation, but they appear to have left matters of cosmology strictly alone. This early example of a flourishing scientific academy rather sticks in the throat of those who overemphasize the relation of science to Protestantism.

Boyle was, when he turned to science, not only young but rich. The son of an enormously wealthy self-made man (the great earl of Cork, an Englishman who made his fortune by exploiting Ireland), he could be and was his own patron. For his type of experimentation, unlike that of Perier, required the expenditure of considerable funds for apparatus and assistants; the effectual pursuit of the advancement of experimental philosophy, Boyle himself pointed out, "requires as well a purse as a brain." Therefore, he argued, people of quality might well "employ the friends of fortune in the search of the mysteries of nature." And the story of Boyle's life shows that he consistently followed his own advice.

THE INVENTION OF THE VACUUM PUMP

One more amateur must be mentioned: the inventor of the vacuum pump, Otto von Guericke. He was a man of affairs, the mayor of Magdeburg who had played an active part in the Thirty Years' War and whose own city was sacked in 1631. His interest in the new experimental philosophy was probably not unconnected with the necessity of his being concerned with matters of military engineering. The story of how he developed his ideas about the atmosphere is obscure; he may have come quite independently to the same conclusions as did Torricelli. He certainly constructed a water barometer and built the first machine for pumping air out of a container. In retrospect his invention is an obvious adaptation of the suction part of a lift water pump. (The retrospect view of all discoveries and inventions must be used with care, lest one fall into the danger of being a "Monday morning quarterback.") Instead of sucking with a piston and cylinder on a column of water, as men had done for centuries in using suction pumps for raising water, von Guericke tried to suck water out of a full wooden cask using a brass pump. There were several models of this invention, and the usual partial success and failure of the pioneer attended his efforts. He attained his results only when he started to pump air as well as water from an enclosed container and finally ended by pumping out air alone. He also found that a spherical metal receptacle was necessary to stand the resulting atmospheric pressure. By 1654 he was able to carry out, before the Imperial Diet assembled at Ratisbon, the famous demonstration of the Magdeburg hemispheres (Fig. 5). Two hollow bronze hemispheres were fitted carefully edge to edge and the air contained in the sphere thus formed removed by a pump. After evacuation the external atmospheric pressure held the two halves together so firmly that a team of eight horses could not pull them apart. Once air was admitted through a stopcock the hemispheres fell assunder.

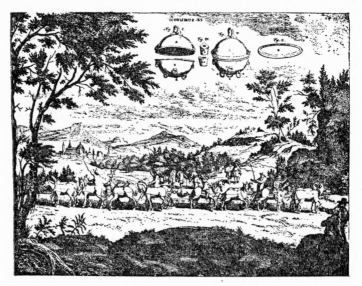

Fig. 5. Otto von Guericke's Magdeburg hemispheres.

This striking demonstration of von Guericke's may be regarded as the verification of another deduction from Torricelli's conceptual scheme. But for our purposes the use of von Guericke's pump by Robert Boyle is more instructive.

ROBERT BOYLE'S EXPERIMENTS

Boyle heard of the new pump through a book published in 1657 by a Jesuit professor at Würzburg (the communication of news of scientific discoveries was still on a most casual basis). Learning of this recent method of producing a vacuum, Boyle saw the possibility of testing still another deduction from the Torricellian conceptual scheme. His combination of logic and imagination represents a pattern repeated by many a successful investigator in the last three hundred years. More than one significant advance in science has come about because someone had the imagination to see that a new instrument made possible the testing of an important point. What Boyle proposed to do was to perform the equivalent of the Puy-de-Dôme experiments in the laboratory. He had modified von Guericke's pump by providing an arrangement by which he could introduce the lower part of the Torricellian barometer into the vessel to be evacuated (Figs. 6, 7, 8). Then, as he worked the pump and withdrew the air from above the mercury reservoir, the mercury column fell. To quote his own words: "All things being thus in readiness, the sucker was drawn down, and immediately upon the egress of a cylinder of air out of the receiver, the quicksilver in the tube did, according to expectations, subside." He was able to cause the column to fall almost but not quite to the level of the mercury in the reservoir. In other words he was able to evacuate the reservoir to a pressure something less than one-thirtieth of the original pressure but, as he suspected, his apparatus was not efficient enough to proceed further with the evacuation. When air was admitted into the receiver the mercury column rose to the usual height.

One would think that by the time Boyle published the account of his experiment the whole learned world would have accepted the new ideas. But the advance of science was slow in the mid-seventeenth century, in part because of the lack of scientific societies and scientific journals. In addition to an elaborate description of his

FIG. 6. Reproduction of a wood engraving of Boyle's first air pump.

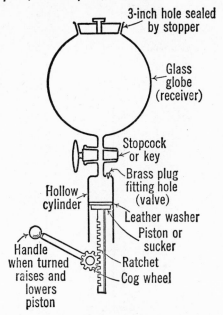

SIMPLIFIED DIAGRAM OF BOYLE'S FIRST AIR PUMP

FIG. 7. The following description of the working of the pump is in Boyle's own words: "Upon the drawing down of the sucker (the valve being shut) the cylindrical space, deserted by the sucker, is left devoid of air; and therefore, upon the turning of the key, the air contained in the receiver rusheth into the emptied cylinder. . . . Upon shutting the receiver by turning the key, if you open the valve, and force up the sucker . . . you will drive out almost a whole cylinder full of air; but at the following exsuctions you will draw less and less of air out of the receiver into the cylinder because there will remain less and less air in the receiver itself. . . ."

pump ("a new pneumatical engine," in Boyle's words), the author described many experiments which could be readily performed in a vacuum. Some of these had been reported by von Guericke and some by the members of the Accademia del Cimento who used an awkward method of producing a vacuum, involving the Torricellian experiment with a tube having an enlarged upper end in which

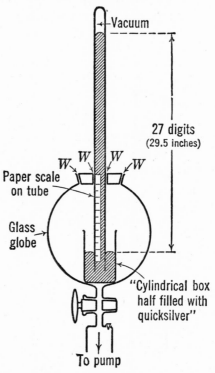

Vacuum

27 digits
(29.5 inches)

W W W W

Paper scale
on tube

Glass
globe

"Cylindrical box
half filled with
quicksilver"

To pump

FIG. 8. Diagram of Boyle's apparatus for removing the air above the reservoir of a barometer. The glass globe is the upper part of the pump. *W* indicates where a sealing wax was used. When the pump is operated the mercury column decreases in height.

equipment could be placed. After the usual inversion this enlarged end became an evacuated chamber (Fig. 9).

To many of Boyle's readers both his experiments and his

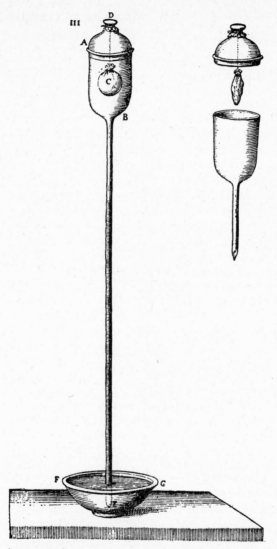

Fɪɢ. 9. Apparatus for an experiment of the Accademia del Cimento. This particular experiment shows the expansion of the air in a bladder when the bladder is placed in a Torricellian vacuum.

ideas were novel. Two at least hastened to attack him. One of these was no less a person than Thomas Hobbes, who though he had belabored the English universities for being old-fashioned strongholds of Aristotelian views was himself a believer in a full universe. The other was an obscure person by the name of Franciscus Linus who had a quaint notion about the phenomenon first observed by Torricelli. Linus had one thing in common with Hobbes—he was a "plenist," that is to say, he believed the universe was full and could not tolerate the idea of a vacuum. Fantastic as Linus' counterproposal seems, it has its appealing side: to explain Boyle's results, Linus put forward the naïve hypothesis of an invisible membrane which held up the mercury column in the Torricellian tube; he called this membrane a *funiculus.* One may note in passing that while this may seem a ludicrous example, it is nonetheless an instance of the use of ad hoc hypothesis, a procedure by no means rare in science. That is to say, faced with an awkward situation, one makes a special postulate to get around it. And in this case Linus could appeal to a common-sense observation to support his hypothetical funiculus. He said in effect: put your finger in the place of the upper end of a barometer and you will *feel* the pull of the funiculus (this can easily be observed with the tube of Fig. 1). What could be more convincing than this argument ad hominem? You feel your flesh being drawn down into the alleged vacuum above the mercury column; of course the same invisible membrane pulls the mercury up!

Boyle replied to this and similar arguments that it is the pressure of the outside air that forces one's flesh into the barometric tube. But he was anxious to answer experiment by experiment. His repetition of the Puy-de-Dôme demonstration with his air pump would not suffice. For Linus answered that in the evacuation of the outside chamber subtle membranes pulled up the mercury in the reservoir and this stretched the funiculus in the barometric tube. Indeed, in the proper mood, you can almost see Linus' funiculus at work like an elastic cord, while the experimenter lowers and raises the column of mercury in the barometer by alternately evacuating the receiver and filling it with air.

Boyle was quick to see that Linus had to make an assumption to account for the approximately constant height of the barometer at sea level. He was forced to postulate that the funiculus could not sustain a column of mercury more than about 29½ inches high. Boyle then proceeded to build a simple apparatus which consisted of a J-tube with one long and one short leg (Fig. 10). He poured in enough mercury to make the difference in levels about 88 inches. Then he sucked with his mouth on the opening, "whereupon (as we expected) the mercury in the tube did notably ascend," to quote Boyle's own words. "Which considerable phenomenon," he goes on to say, "cannot be ascribed to our examiner's Funiculus, since by his own confession that cannot pull up the mercury, if the mercurial cylinder be above 29 or 30 inches of mercury."

From these experiments designed to combat the strange notions of Linus came, as a by-product, quantitative measurements that resulted in the formulation of the famous Boyle's Law relating the volume and pressure of a gas. But a consideration of this phase of Boyle's work will be reserved for a later chapter. The present discussion of seventeenth-century pneumatics may well close with a consideration of some ingenious experiments by which Boyle sought in vain to obtain evidence of the existence of that "subtle fluid" which had been postulated by Descartes and was an article in the creed of all plenists. The case is of special interest because almost all the "case histories" which can be studied by the layman involve the verification of deductions from conceptual schemes which have proved fruitful and therefore had longevity. The so-called

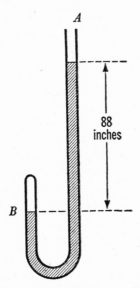

FIG. 10. When Boyle sucked with his mouth at A, the mercury rose in the long arm of the J. If a funiculus from the tip of his tongue was dragging up the mercury, it must be supporting a column of mercury 88 inches high. This was contrary to Linus' hypothesis that the funiculus was an invisible membrane that would support a column of mercury about 30 inches high and no higher.

negative results, the failures if you will, are less readily available.

Hobbes followed Descartes in believing the universe was full of a subtle fluid. He carried over into this new notion the older Aristotelian explanation as to why a wine barrel needs a hole in the top ("to let something in") if the liquid is to come out. He seems to have been willing to agree that Boyle had pumped something out of his receiver but refused to admit the existence of a "real vacuum."

Boyle on this point (as on many others) was cautious. In his first report he raised the question of whether his experiment with the evacuation of a receiver proved that the space was "truly empty, that is devoid of all corporeal substance," and pointed out the difficulties presented to those who answered the questions in the negative or the affirmative. For, to quote his own words, "on the one side it appears that notwithstanding the exsuction of the air, our receiver may not be destitute of all bodies, since any thing placed in it, may be seen there; which would not be, if it were not pervious to beams of light . . . and either these beams of light are corporeal emanations from some lucid body or else at least the light they convey doth result from the brisk motion of some subtle matter. . . ." (One may remark parenthetically that fifty years ago these two alternatives would have been recognized as mutually exclusive choices for an adequate theory of light, but not today, as was noted in a previous chapter.) "On the other side, it may be said that as for the subtle matter which makes objects visible . . . it may be presumed to pass through [the glass walls of the receiver] . . ." (This presumption would have probably been stated as correct in the 1890's.) Having weighed the pros and cons, Boyle states: "Nor dare I yet take upon me to determine so difficult a controversy."

With his first air pump Boyle could hardly have tackled the difficult problem of searching for a subtle fluid. But the young English experimenter had soon become dissatisfied with his air pump and had a still better engine designed. The second model had a separate receiver which could be evacuated (Fig. 11). With this contrivance and by the in-

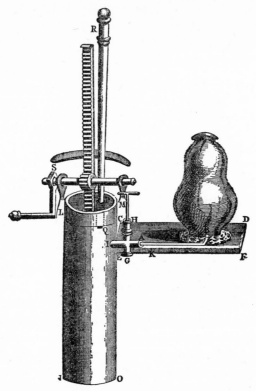

Fig. 11. A partially diagrammatic illustration of the upper portion of Boyle's second air pump. The lower portion of this pump was essentially the same as in the first model. The plate *CDEF* is cut away to show the tube *AB* through which air is pumped out from the glass "receiver" via the valve *HG*.

vention of a gauge to measure the residual pressure in the evacuated vessel, Boyle was well on his way toward modern experimental methods. Indeed, I venture the opinion that Robert Boyle is the real father of experimental science.

For he was not only an ingenious and careful investigator but the first to set standards of complete and accurate reporting.

Boyle published in 1667 a long report on many experiments with the second model of his engine. Some of these will be considered in the next chapter. Here we are concerned only with several which were entitled "An attempt to examine the motions and sensibility of the Cartesian Materia subtilis or the Æther. . . ."

Boyle set out in a deliberate manner to examine a hypothesis on a grand scale which we may or may not choose to designate a conceptual scheme. It is worth noting that we have the same difficulty in nomenclature as regards the luminiferous ether of the nineteenth century. Indeed, we can loosely identify the two and call the notion of a subtle fluid a conceptual scheme that still has its uses at least for pedagogic purposes in the twentieth century.

Neither Boyle nor anyone else could test the hypothesis of the existence of a subtle fluid *directly* any more than one could test the Torricellian scheme directly. This is a highly important point. Consequences must be first deduced from a conceptual scheme (p. 51) and then these consequences in turn become the basis for a chain of reasoning terminating in a very limited working hypothesis. It is the limited working hypothesis in the last analysis that a manipulation followed by an observation will or will not verify. Boyle does not analyze his results or his procedures in such a way as to make clear the distinction between testing the existence of a subtle fluid in general and the presence of a fluid with certain properties. But the difference (which is essential) is implicit in the conclusion of one of his experiments. Here he states explicitly that if there is an "æther"

in his evacuated vessels it must be "thinner" than common
air rarefied a hundred times. With this limited conclusion
one can hardly quarrel.

What Boyle did in effect was this: he postulated the existence of a fluid with certain properties defined in terms of
experiments he envisaged. He then deduced from his version of a more general and vague conceptual scheme certain consequences; these in turn led to the "if . . . then"
type of reasoning and to a series of specific experiments. In
every case the results were negative; that is to say, the predicted effects were not observed. The accumulation of
negative answers to specific experimental tests made improbable the existence of a subtle fluid with the particular
properties postulated: they showed nothing, of course, as
to the existence of a subtle fluid with other properties.

Boyle's experiments defined a subtle fluid as a fluid
strictly comparable to common air under a pressure of
something less than a thirtieth of atmospheric pressure.
When thus attenuated, common air can still be moved by
the quick motion of a bellows or a syringe, and the existence of a blast of such a "thin" fluid can be made manifest;
Boyle showed that even after the pressure had been reduced below an inch of mercury the air in his receiver could
be blown against a feather with sufficient force to make the
latter move (Fig. 12). Boyle had to assume that a subtle
fluid such as he sought would either not be removed from
his receiver by the operation of his pump or would immediately leak back. For one of the postulated properties of this
fluid was the ability to leak back through his valves and
through the openings he had sealed with wax. He had some
reason for suspecting the existence of such a fluid, for like
all experimenters with vacuum equipment Boyle was
plagued by "leaks." In his first paper he indeed suggests

that perhaps some portion of common air was so subtle as to leak through orifices he thought were tight.

The reader may be inclined to dismiss as childish Boyle's experimental attempts to find evidence for Descartes's ether. But they are a good illustration of the way a series of well-executed experiments that give negative answers

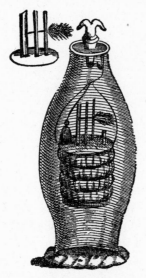

Fig. 12. Boyle's picture of the "bladder-bellows," used in trying to find a medium more subtle than air.

may make improbable a conceptual scheme. They have little bearing on the relevance of a general speculative idea, but they do provide evidence about a conceptual scheme provided definite postulates are introduced that are directly related to the experiments in question. In terms of the conceptual schemes which all scientists today take for granted, Boyle's experiments seem naïve. But we must remember that for all the seventeenth-century investigators

knew, air might have been composed of two or more materials differing in their ability to pass through very fine
holes; such a difference is taken advantage of whenever we
strain out a finely divided solid from a liquid, for example.
Indeed, we now know that there is a very slight difference
in the rate at which the constituent gases of the atmosphere
(chiefly oxygen and nitrogen) flow through a tube of very
small diameter. But this difference is so slight that it is not
reflected in any behavior of air in experiments which could
be performed with the equipment available in the seventeenth or even the eighteenth century. Indeed, we now
believe that the great difference in "subtlety" for which
Boyle was looking cannot exist in any mixture of gases.
It is in the nature of a gas that there can be no gross non-
homogeneity in the mixture such as occurs in a suspension
of fine clay in water or even in water solutions of the materials which are present in blood or milk. More than a century elapsed, however, before it became obvious that such
was indeed the case. And it was a century or more before
the conceptual scheme was developed which we now use in
our explanations of the behavior of air and other gases (the
kinetic theory of gases).

In the presentation of seventeenth-century ideas about
air and atmospheric pressure I have barely alluded to the
fact that air, unlike water, is easily compressed. No one
needs to be reminded of this today. Interestingly enough,
however, little attention was paid to this property of common air until after Boyle published the account of his experiments. To be sure, Torricelli in one of his famous letters to Cardinal Ricci uses the analogy of a cylinder of
wool as illustrating how a mass of compressible material
exerts pressure on a supporting surface, but his method of
producing a vacuum could be explained without mention-

ing the great compressibility of air. Pascal in his writings stresses the analogy between water pressure and air pressure and, though alluding to the wool analogy, treats air for the most part as though it were merely a very much less dense medium than water. Boyle, on the other hand, emphasizes repeatedly the importance of the compressibility of the air. Indeed, he uses the picturesque phrase, "the spring of the air."

If one is looking for a simple example of "a concept arising from experiment," the rebirth of the idea of air as an elastic fluid is made to order. For Boyle's method of producing a vacuum, unlike the method used by the Florentines or Pascal, involved the use of a pump. And one has only to operate the handle of an air pump today to feel the "spring of the air"! If the pump is for compression (for blowing up tires) you feel as if you were pushing against a spring on the downstroke, if it is a pump for producing a vacuum you have the same sensation on the upstroke. Indeed, an examination of a pump for producing a vacuum on the von Guericke principle (Fig. 7) shows that if air were as incompressible as water, the machine would never work. The essentially instantaneous expansion of the air in the receiver assures its distribution into the evacuated cylinder after the piston has been drawn down. On each stroke a fraction of the remaining air is thus removed (the fraction depends on the relative size of the receiver and the cylinder). Boyle, the great experimentalist of the seventeenth century, cannot be given absolute priority for the concept of the compressibility of air, but he probably evolved it independently and certainly was the first to realize its importance. As a first approximation we may say air is easily compressible and water almost incompressible. These qualitative concepts were useful in the first stages of the development of a

new branch of science. Before very long, however, they
proved inadequate unless quantitative statements could be
added. To do this requires measurements and the manipulation of abstract ideas by means of mathematics. How
Boyle's qualitative concept of the spring of the air developed into a quantitative formulation of the elasticity of
a gas forms a portion of a subsequent chapter (Chapter
VI) on quantitative experimentation and the role of mathematics.

CHAPTER FIVE

Some Recurring Patterns in
Experimental Inquiry

I N THE preceding chapter several illustrations were provided of one recurring pattern in experimental science,
namely, how the consequences of a new conceptual scheme
may be tested by experiment. These examples likewise
demonstrated how a new conceptual scheme may be fruitful of new experiments. The Torricellian experiment, Perier's ascent of the Puy-de-Dôme, Boyle's repetition of
Perier's observation in the laboratory, all these are classic
instances of these two recurring patterns. In all these the
new conceptual scheme was fruitful; furthermore, the results of the experiments tended to confirm the postulate
of a hypothesis on a grand scale. On the other hand, Boyle's
search for a certain kind of subtle fluid (p. 92) yielded only
negative results; consequently the work soon came to a
dead end. The concept of a subtle fluid of this type must be
judged, therefore, to have been essentially fruitless, and
except in our most cautious and skeptical mood we would

label it as "incorrect." At the minimum we may say that none of the consequences derived from the concept could be confirmed, and a scheme which yields only a few experiments the results of which are negative is certainly not by any standard fruitful.

In this chapter we shall consider some elementary examples of certain other recurring patterns in experimentation. For this purpose the study of seventeenth-century pneumatics is rewarding. Many significant aspects of the tactics and strategy of science are here well illustrated, though there is one serious omission. It is difficult to provide an example of one highly important point, namely, how a new conceptual scheme may be developed from the first flash of genius, for unfortunately we know very little about the origin of Torricelli's broad hypothesis which so soon became a new conceptual scheme. But this deficiency I hope to remedy in Chapter VII which deals with the chemical revolution. Otherwise, almost all the essential elements of the use of the experimental method in advancing science can be found by analyzing the work of Robert Boyle and his contemporaries. Boyle's placing the "spring of the air" in the forefront of his discussion (p. 95) illustrates how a new concept arises from experiment. The significance of the invention and improvement of instruments stands out clearly in this period of scientific history. One has only to mention von Guericke's pump, the several models of Boyle's "pneumatical engine," and Torricelli's barometer to realize how fundamental new devices are in opening new fields for experimentation. Yet by themselves experimental observations hardly constitute an advance in science. A chain of argument that links the experiment to a broad idea (a new concept or conceptual scheme) is an essential element; and such chains are nicely exemplified by some of the simple yet classic experiments initiated by Pascal and Boyle and considered in the preceding chapter.

The early history of pneumatics affords many instances which can be cited to show the supreme importance of the recognition of the variable factors in an experiment. One could open almost at random Boyle's accounts of his experiments "touching the spring of the air" and find a text for a sermon on the control of variables. His study of the transmission of sound in a vacuum is worth brief consideration for this reason. It may also serve to remind the reader of one type of experimentation which has played a useful part in the forward movement of scientific inquiry: an investigator has at hand a new or improved device and realizes that it can be used to make more certain a somewhat ambiguous interpretation of past experiments. The chances of a revolutionary discovery are slight; the concepts or conceptual schemes involved are quite generally accepted. Still there are some loose ends that need to be tied down; in short, the matter while not of the highest urgency seems well worth studying because a new line of attack can be envisioned with the instrument at hand. Converging evidence, we must remember, provides the verdicts on which the fabric of science rests. What follows in the next few pages may seem extremely trivial to some readers. Yet for that reason it may serve a useful purpose. For one danger in presenting science to laymen through the case history method is that it tends to leave the impression that only wide conceptual schemes or radically new concepts are of significance. E. S. Creasy's *Fifteen Decisive Battles of the World* creates something of this illusion in the mind of the average reader as to military history. The next few pages, though designed to illustrate several points in the tactics of science, may introduce an antidote to any grandiose view of scientific history, for I shall review some relatively unimportant experiments performed by Boyle with his air pump.

The Transmission of Sound in a Vacuum. If we may judge from Boyle's discussion of the subject, it had been generally believed for some time that air was the medium through which sound was transmitted. From this concept it followed that sound should not pass through an evacuated space. The Florentine experimenters of the Accademia del Cimento seem to have attempted to test this consequence of the concept of air as the medium for the transmission of sound. Their results were inconclusive, which is not strange considering their method of operating. For what they did was to suspend a bell in the enlarged top of a barometer tube, a cumbersome contrivance at best! Boyle's pump obviously provided a much more convenient method of producing a vacuum. Still with the first model (Figs. 6, 7) his results were likewise ambiguous. A watch was suspended by a thread inside the globular receiver. The ticking could be plainly heard when the glass globe was full of air but not when it was evacuated. On the other hand, a bell suspended from a stick pressing against the sides of a receiver could be heard almost as well when the receiver was evacuated as when it was full of air.

Boyle recognized two possible sources of error. There might be some air still left in the evacuated space (bubbles of air easily get in the mercury column the Florentines used); the path of the sound might be through the solid support to the wall of the receiver and thence to the

outside air. When his second model pump with the separate receiver (Fig. 11) was ready, more effective experimentation became possible. A "watch with a good alarm"
was suspended by a thread in the receiver which was then
evacuated. When the time came for the alarm to ring nothing could be heard. But immediately after some air was

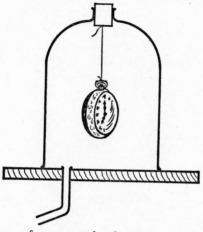

Fig. 13. Diagram of apparatus for the experiment with a watch in
a vacuum. The watch is suspended in a bell-shaped glass "receiver"
from which the air is pumped out.

admitted, a faint sound was heard; on the admission of
more air "the by-standers could plainly hear the noise of
the alarm" which was still ringing. (Fig. 13).

All these results were convincing evidence but not sufficient to suit Boyle. He went back to the matter of silenc-
ing a bell in vacuo. With an apparatus shown in Fig. 14 he
was able to sound a bell suspended inside the receiver by
a thin bent wire. On turning the "key" in the top of the
receiver, a hammer was made to strike the bell. "When the

receiver was well emptied," Boyle reports, "it sometimes seemed doubtful, especially to some of the by-standers, whether any sound was produced or no; . . . when a little air was let in, the stroke of the hammer upon the bell (that before could now and then be not heard, and for the most

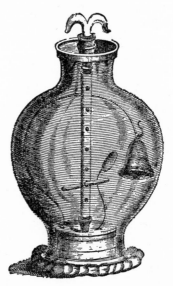

Fɪɢ. 14. Reproduction of a wood engraving of Boyle's apparatus for striking a bell in a vacuum.

part be but very scarcely heard) began to be easily heard; and when a little more air was let in, the sound grew more and more audible . . ."

The two variables in these simple experiments were the possible presence of air and the transmission of sound through a solid support. The recognition of these variables led to their control by elimination. Of the first, it is interesting to note that Boyle later perfected a gauge for measuring the pressure inside his evacuated receiver. If he had had such a measuring instrument at hand initially, he could have made more precise the control of the first variable just mentioned, namely the degree of exhaustion of the receiver. He would then have been able to say something to this effect: "When the air pressure was reduced to less than a certain reading of the gauge, no sound could be heard when the alarm rang (or the bell was struck); when air was admitted and the pressure rose to a definite value, then a slight sound was heard." *When the quantitative measurement of a variable becomes possible, the uncertainty introduced by the variable is greatly reduced and experimentation is usually enormously simplified.*

Exploration with a New Technique. We cannot leave Boyle and his air pump without drawing another moral from the tale. If one turns the pages of his rather long-winded account of his many experiments one cannot fail to be impressed by what seems mere random experimentation. And that this type of activity has played a vital part in the history of physics and chemistry no one can well deny. There is no doubt about it, the logically coherent pattern which has been the subject of our analysis hitherto is in such instances conspicuous by its absence. The inventor of a new scientific tool behaves like an explorer landed on a hitherto unmapped island. He will proceed to make the most of his good fortune, and every observation appears to be worth making and recording. Boyle, for example, kept putting to himself this sort of question: what will happen if I place this or that object in a vacuum? Before his time, the members of the Accademia del Cimento and Otto von Guericke had similarly explored "phenomena in vacuo." Probably most of their results were unknown to Boyle but we need not trouble in this account with the matter of priority. What is important is to recognize the avid experimentalist making the most of the potentialities of his new technique. Sometimes he will be testing the consequence of a broad conceptual scheme as was the case, discussed in the last chapter, where Boyle was experimenting with a barometer in vacuo. Sometimes he will attempt to present new evidence for a well-recognized idea, as in the instance just discussed of the experiments on sound. But often he will be *just experimenting,* and his discoveries may stand as isolated bits of information to be woven into the conceptual fabric of science at a much later date.

The mention of a few of Boyle's observations may serve to illustrate both the exploratory nature of certain kinds of experimental activity and the often fragmentary nature of the results. He demonstrated with the aid of some ingenious contrivances the fact that a candle would not burn in a vacuum but gunpowder would. It was not until the late eighteenth century, however, that a satisfactory conceptual scheme was developed that could incorporate these simple facts. The failure of ordinary combustion in a vacuum and the death of small animals in an evacuated receiver certainly indicated that air was essential for burning and life. But as we shall see later it is a long step from this to the discovery of oxygen. Indeed, an apparently devious path had first to be followed, a path that led through some strange notions about a mysterious substance known as phlogiston.

In some instances Boyle was aiming his inquiry at a specific target. For example, he went to great pains to devise a method of rotating two surfaces against each other in an evacuated vessel. He then quickly admitted air and found the objects to be warm. Thus he concluded that the generation of heat by friction would take place in the absence of air. With his first pump he studied what happened when a vessel filled with water but open at the top was subjected to a diminished pressure of air. He seems at the outset to have desired to learn whether the water would expand in volume appreciably when the atmospheric pres-

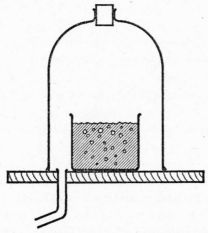

FIG. 15. Diagram of apparatus for boiling water in a vacuum. When air is pumped out of the bell-shaped glass "receiver" the water begins to boil.

sure was greatly reduced. He at once ran into trouble, however, because ordinary water contains dissolved air. Bubbles of this air at once appear and rise to the surface when the air pressure above a sample of such water is diminished. This complicates the problem. And at first Boyle was not sure whether there was an appreciable "spring" to water or not (i.e., whether it was appreciably expansible). It is easy to confuse the rising of air bubbles with the true boiling which takes place when warm water is placed in even a poorly evacuated container. Only later experimentation with his improved pumps led Boyle to recognize that water at room temperature can be made to boil if the air pressure is reduced below about a thirtieth of atmospheric pressure (Fig. 15). Again it was a hundred years before the full implications of this and similar observations could be formulated by means of a series of concepts involving the idea of vapor pressure.

Boyle's Contributions to Techniques of Experimentation. Boyle kept on with his experiments with air pumps for a considerable portion of his life, though he is perhaps more famous for his contributions to chemistry, which have to some degree been overpraised. To his volume entitled *The Skeptical Chymist* we shall refer later. The third model of his air pump was built in 1669 in collaboration with an inventive Frenchman, Denis Papin. With it Boyle was able to produce a vacuum in which the pressure was considerably less than one-hundredth of atmospheric pressure. He then showed how in such an evacuated vessel one could produce artificial airs by such devices as placing coral in acid (he really prepared carbon dioxide). With ingenious and elaborate methods he could then transfer this artificial air from one container to another and thus experiment with it. Boyle likewise showed how, by using diminished pressure, liquids could be distilled at temperatures much below the normal boiling point. The striking part of the history of pneumatics is that these new techniques made little impression on the course of science. Only in the nineteenth century did "distillation in vacuo" come into its own as a procedure regularly employed by chemists. And only in the twentieth century did experimenters with gases (artificial airs) depend on the use of evacuated vessels.

Why was there this long lag in the adoption by the scientific world of Boyle's advances in the technique of pneumatic investigation? The answer seems to lie in the difficulties of this manner of experimentation. Pumps like Boyle's were extremely expensive; vacuum technique is a fussy art and has remained so even to the present day. A far more convenient but cruder method of handling gases was invented about the same time and received almost universal acceptance. This was the use of the pneumatic trough about which we shall have much to report in connection with the chemical revolution. Not until glass blowing and metalworking had reached a high stage of development could laboratory workers rapidly construct apparatus for handling gases along the lines suggested by Boyle. And it was not until satisfactory and relatively cheap vacuum pumps were available that considerable contamination with residual air could be avoided. In the second half of the nineteenth century the development of the first incandescent light stimulated inventors to produce pumps that would reduce the pressure to less than several hundredths of atmospheric pressure. Today such pumps make it possible to evacuate even large vessels to a pressure well below a millionth of an atmosphere. X-ray tubes, radio tubes, cyclotrons, and all manner of complicated physical and chemical equipment are possible because the art of "producing a good vacuum" has become a commonplace affair. The laboratory consequences of Boyle's arduous experimentation have finally borne fruit.

One more word is in order before we finally leave the subject of Robert Boyle and his pumps. Denis Papin, his collaborator, has achieved a certain degree of fame as the inventor of "Papin's digestor." This is none other than the pressure cooker. How the idea originated seems quite clear.

Boyle and Papin studied not only the behavior of material in vacuo (including such foods as grapes) but also the behavior of substances in compressed air. The effect of pressure in increasing the temperature of boiling water was thus brought to their attention. It is interesting that Papin's digestor though mentioned off and on in the scientific literature did not come into its own as a practical device until a few years ago. This modern addition to the kitchen, prized highly by the housewife, was nevertheless used for cooking purposes at the time of its invention. John Evelyn in his famous diary under date of April 15, 1682, records with appreciation that the members of the Royal Society partook of a supper cooked in Pepin's digestor. He remarks, "This philosophical supper caused much mirth among us and exceedingly pleased all the company."

THE ROLE OF THE ACCIDENT

Science is sometimes presented as though it were exclusively the work of high-powered mathematicians elaborating theories and sometimes as though it were all a matter of blind chance. As a consequence the reader is often confused in regard to the role of what appears to be an accidental observation. This is particularly true in connection with the development of new techniques and the evolution of new concepts from experiment. The case history which I recommend for a study of these subjects is the work of Galvani and Volta on the electric current. This case illustrates the fact that a chance observation may lead by a series of experiments (which must be well planned) to a new technique or a new concept or both. It also shows that in the exploration of a new phenomenon the experiments may be well planned without any "working hypothesis" as to the nature of the phenomenon but that shortly an explanation is sure to arise. A new conceptual scheme will then be evolved. This may be on a grand scale and have wide applicability or may be strictly limited to the phenomenon in question. A test of the new concept or group of concepts in either instance will probably lead to new discoveries and the eventual establishment, modification, or overthrow of the conceptual scheme in question.

Galvani's Discoveries. The story begins with certain observations made by Luigi Galvani, an Italian physician and professor at Bologna, some time before 1786. This investigator noted the twitching of a frog's leg when the crural nerves were touched by a metallic scalpel in the neighborhood of an electrostatic machine from which sparks were drawn. *He followed up his observation:* this is the significant feature of the history of this episode. Time and time again throughout the advance of science the consequences of following up or not following up accidental discoveries have been very great. The analogy of a general's taking advantage of an enemy's error or a lucky break is clear. Pasteur once wrote that "chance favors only the prepared mind." This is excellently illustrated by the case at hand. The Dutch naturalist Swammerdam had previously discovered that if you lay bare the muscle of a frog in much the same way as Galvani did, grasp a tendon in one hand and touch the frog's nerve with a scalpel held in the other hand, a twitching will result. But Swammerdam never followed up his discovery. Galvani did. In his own words, "I had dissected and prepared a frog . . . and while I was attending to something else, I laid it on a table on which stood an electrical machine at some distance . . . Now when one of the persons who were present touched accidentally and lightly the inner crural nerves of the frog with the point of a scalpel all the muscles of the legs seemed to contract again and again. . . . Another one who was there, who was helping us in electrical researches, thought that he had noticed that the action was excited when a spark was discharged from the conductor of the machine. Being astonished by this new phenomenon he called my attention to it, who at that time had something else in mind and was deep in thought. Whereupon I was inflamed with an incredible zeal and eagerness to test the same and to bring to light what was concealed in it." [*]

Galvani did not succeed in bringing to light all that was concealed in the new phenomenon. But he proceeded far enough to make the subsequent discoveries inevitable. In a series of well-planned experiments he explored the obvious variables but without having a clear-cut over-all hypothesis. This is the usual situation when a new and totally unexpected phenomenon is encountered by a gifted experimenter. A series of working hypotheses spring to mind, are tested, and either discarded or incorporated into a conceptual scheme which gradually develops. For example, Galvani first determined whether or not sparks had to be drawn from the electrical machine in order to occasion twitching. He found "Without fail there occurred lively contractions . . . at the same instant as that in which the spark jumped. . . ."

The nerves and muscles of the frog's leg constituted a sensitive detector of an electric charge. Galvani found that not only must a spark be passing from the electrostatic machine but the metallic blade of the scalpel must be in contact with the hand of the experimenter. In this way a small charge originating from the electrical disturbance, namely the spark, passed down the conducting human body

[*] Reprinted by permission from W. F. Magie, *A Source Book in Physics* (McGraw-Hill Book Company, Inc., 1935).

through the scalpel to the nerve. So far the physician was on sound and fertile ground. There now occurred one of those coincidences which more than once have baffled an investigator initially but eventually have led to great advances. The frog's leg could under certain circumstances act not only as a sensitive electrical detector but as a source of electricity as well. When this happened the electricity self-generated, so to speak, actuated the detector. One can readily see that the superposition of these two effects could be most bewildering and misleading. This was particularly so since the conditions under which the frog's leg became a source of electricity were totally unconnected with any electrical phenomena then known. The variable was the nature of the metal, or rather metals, used. For Galvani discovered and duly recorded that the electrostatic machine could be dispensed with if the leg and the nerve were connected by two *different* metals. Under these conditions the twitching occurred. (The experiment was usually performed as follows: a curved rod was made to touch simultaneously both a hook passing through the spinal cord of the frog and the "muscles of the leg or the feet.") "Thus, for example," wrote Galvani, "if the whole rod was iron or the hook was iron . . . the contractions either did not occur or were very small. But if one of them was iron and the other brass, or better if it was silver (silver seems to us the best of all the metals for conducting animal electricity) there occur repeated and much greater and more prolonged contractions."

Galvani had discovered the principle of the electric battery without knowing it. His two metals separated by the moist animal tissue were a battery, the frog's leg the detector. Every reader can perform the equivalent of Galvani's experiment himself. A copper coin and a silver one placed above and below the tongue when touched together produce in the tongue a peculiar "taste." A very small electric current flows, and our tongue records the fact through a series of interactions of electricity and nerves much in the same way as did Galvani's "prepared" frogs. Not having a suspicion of all this, however, Galvani developed a hypothesis on the grand scale to account for all the phenomena in terms of what was then known about electricity, which was derived entirely from experiment with electrostatic machines. Now that outside electrical disturbances had been found unnecessary (when he unwittingly used the *right* metallic combination!), the results, he says, "cause us to think that possibly the electricity was present in the animal itself." Galvani's following up of an accidental discovery by a series of controlled experiments led to a recording of the significant facts, but it was to be another Italian who developed the fruitful concept. It was Volta who in the late 1790's, continuing the study of the production of electricity by the contact of two different metals, invented the electric battery as a source of what we now often call Galvanic electricity.

Volta's Invention of the Electric Battery. Alessandro Volta of Padua had earlier invented a new form of instrument for detecting small charges of electricity. He began by agreeing with Galvani about animal electricity and went about studying it. With his new instrument, a sensitive condensing electrometer, Volta explored various combinations of variables related to Galvani's early experiments and found that the frog could be eliminated in favor of almost any moist material. This discovery might be considered an example of an accidental observation, but if so it is of a different order from that of Galvani. Explorations with new techniques and tools, as we have already noted, if undertaken in a more or less orderly fashion, very often turn up unexpected facts. In this sense a great majority of new facts of science might be called accidental discoveries. Yet the difference between this sort of experience and the example afforded by Galvani's work is clear. Galvani as a doctor and anatomist was interested in muscles and their action, not in electricity. It was a fortuitous circumstance that an electrical machine became connected with the first observations. But all the more credit goes to Galvani for following up the fortunate accident since what he noted lay so far outside his main interests.

Volta's new discovery amounted to the invention of the electric battery; for he showed that electricity was produced when two different metals were separated by water containing salt or lye. This was most conveniently done by using moistened paper. In a letter to the president of the Royal Society of London in 1800 Volta wrote: "30, 40, 60 or more pieces of copper, or rather of silver, each in contact with a piece of tin, or of zinc, which is much better, and as many layers of water or of some other liquid which is a better conductor than pure water, such as salt-water or lye and so forth, or pieces of paste-board or of leather, etc. well soaked with these liquids; . . . [Fig. 16] such an alternative series of these three sorts of conductors always in the same order, constitutes my new instrument; which imitates . . . the effects of Leyden jars. . . ." This new battery was a source of electricity different from the electrostatic generator already known in 1800; it was the first source of steady current, whereas the sparks from a frictional machine are but brief spasms of current.

There was a hot controversy between Galvani's disciples (Galvani died in 1798) and Volta about whether or not there was such a thing as animal electricity and what caused the twitching of the frog's leg in the first experiments. Volta soon lost interest in the quarrel and devoted his attention to the study of his new battery. Today we have a rather complete and highly satisfactory conceptual scheme in which the facts about electric batteries find their place. This is not the case, however, with observations about muscles, nerves, and electric currents in ani-

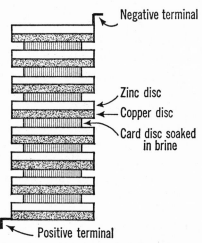

Fɪɢ. 16. Diagram of one form of Volta's battery or pile.

mal tissue. In this field one working hypothesis still replaces another and new experiments are continually throwing new light on old observations. In a sense, we have not yet finished with Galvani's very first experiment but have finished with Volta's discovery. The original controversy centered on the question, Is there animal electricity? This has now become largely a meaningless question, but in attempting to find an answer Volta discovered the electric battery. Such is often the course of scientific history. We end by solving a problem other than the one first at issue.

The Discovery of X Rays. A nineteenth-century episode which illustrates how well-planned experiments may be used to follow up an observation is furnished by the discovery of X rays. The story is familiar to all scientists, though perhaps it is not generally known that before Roentgen announced his discovery several other investigators noticed the fogging of photographic plates near an electric discharge tube. Roentgen followed up his observation; the others did not. But the clue from which Roentgen worked can hardly be considered a mere happy accident. For Roentgen was studying the stream of electrons (they were then called simply cathode rays) which can pass through a thin window in an electric discharge tube. He was aware that these rays would cause fluorescence of certain substances. He consequently had at hand a screen coated with such a substance and observed that it shone even when it lay at some distance from the tube. Following up this observation Roentgen quickly demonstrated that some sort of radiation which passed through not only glass but opaque substances was responsible for the effect. From then on he was able to devise better methods of producing these rays and thus introduced a revolutionary technique.

The Discovery of the Rare Gases. One of the most striking instances of an investigator's "turning the unexpected corner" is the discovery of the presence of the rare gases in the atmosphere. In any logical presentation of science this case should be considered after a full examination of quantitative experimentation and after some consideration of chemical phenomena. But since it illustrates the pattern of inquiry with which we have just been concerned I am venturing to present the case in grossly simplified form as the concluding section of this chapter.

We start with a consideration of the troubles of a physicist and end with a chemical discovery. Lord Rayleigh, a physicist, had devoted some twelve years of active experimentation to the careful determination of the relative densities of the gaseous elements. This is a much more difficult matter than it sounds. Rayleigh wanted results reliable to within one part in ten thousand, and this required elaborate experimental precautions both of a chemical and physical nature. Why it seemed important to this physicist, in the latter part of the nineteenth century, to measure the relative weights of equal volumes of the gaseous elements is another story. For our present purposes we need only concentrate attention on the series of events which led to Rayleigh's arduous labors. In 1892 in a note published in the weekly journal *Nature* Rayleigh wrote that he was "much puzzled by some recent results as to the density of nitrogen" and stated he would be obliged "if any of your chemical readers can offer suggestions as to the cause."

Air is now known to be a mixture of the elements nitrogen, oxygen, and argon, with traces of other substances. In 1890 air was believed to consist of only nitrogen and oxygen, and Rayleigh therefore believed he could prepare nitrogen by removing the oxygen from the air. The puzzle was this: one particular method appeared to yield a nitrogen that was slightly heavier, volume for volume, than nitrogen prepared by removing oxygen by other procedures. The difference was only one part in a thousand, but the results were quite consistent. Rayleigh had refined his physical methods so that the density of nitrogen, as determined from different samples prepared by the same method, varied only by about one part in ten thousand. In other words the discrepancy was ten times as large as what might be called the uncertainty of the measurement. The question was: To what is the discrepancy due?

The question, while puzzling, could easily have been pushed aside. But Rayleigh was not content to do so. *He followed up this observation.* Two years later he announced in a paper in the *Proceedings of the Royal Society* that the puzzle not only remained but matters had got much worse. (A veteran experimenter I once knew used to love to exclaim in the midst of a problem, "Things have to get much worse before they will get better," and he was usually right!) Nitrogen prepared from air was actually heavier than nitrogen prepared from any of its compounds, such as ammonia, by about one part in two hundred! The reason why the difference first noted was so slight (one part in a thousand) at least was clear. The method of removing oxygen from air had involved the use of ammonia, and *some* of the nitrogen in the sample came from the ammonia, not from the air.

The situation was now something of a scientific scandal. Here it was the end of the nineteenth century and everyone thought they knew all about the simple elements and certainly about common air. (The idea of isotopes was still in the future by nearly twenty years.) And yet two ways of preparing samples of an element yielded different materials as judged by their densities. The unexpected corner had really been turned; starting with but a faint clue Rayleigh had shown that there was a real and quite unimagined problem almost on the doorstep of every chemist. From here on it was only a matter of time before the answer to the puzzle would be found. The answer was simple: nitrogen prepared from air by removing the oxygen is not pure nitrogen; it contains argon, a heavier gas, in considerable amounts and traces of the other rare gases as well. These elements are not removed by any of the procedures used for eliminating oxygen.

It was with rather red faces that the chemists had to admit at the turn of the twentieth century that for a hundred years they had missed as much as 0.5 per cent of a gas present in the air we breathe. Yet there was some professional satisfaction in the fact that a chemist, Sir William Ramsay, shared the honors of the discovery with Rayleigh. Working at first independently and then jointly these two men proceeded to isolate the heavy constituent of the atmosphere—essentially argon—by removing both the oxygen and nitrogen. Ramsay used a method which depended on the formation of a compound between the element nitrogen and the metal magnesium; this was something that could not have been accomplished a few decades earlier since magnesium became available only in the late nineteenth century. Rayleigh, however, reverted to an observation of more than a century before. Henry Cavendish in the 1780's had reported that he had succeeded in combining nitrogen and oxygen (to use modern words) by passing an electric spark through a mixture of the two gases. Since the compound thus formed was soluble in water, he had at hand a method of determining the homogeneity of a sample of atmospheric nitrogen (again using modern terminology). He set out to do exactly this and reported that if there is any part of the nitrogen of our atmosphere which differs from the rest ". . . we may safely conclude that it is not more than $\frac{1}{120}$ of the whole." This figure was no guess on the part of Cavendish. He had actually seen and measured a residual gas which was not absorbed by his procedure and this small bubble amounting to about 1 per cent of the nitrogen was undoubtedly argon. Yet no one had ever bothered to follow up this lead and examine the nature of what was left. Hundreds of chemists, at least, must have read those words of Cavendish over the years, and they all missed a chance to make a great discovery, probably thinking that the small bubble merely represented the failure of Cavendish's procedure, that he had not succeeded in using up all the nitrogen.

Rayleigh repeated Cavendish's experiments and thus isolated argon. The new gas (new, that is to the scientific world), whether isolated by the Ramsay method or the Cavendish-Rayleigh method, had unusual properties. Its existence and that of its companion rare gases altered the view of chemists about many fundamental matters. In short, this was a discovery of the first order of importance for it opened up many more fields of inquiry both experimental and theoretical. To be sure, some of these fields would not have been ready for cultivation twenty-five or fifty years earlier. Indeed, two of the instruments which proved invaluable in characterizing argon and the other rare gases—the electric discharge tube and the spectroscope—would not have been at hand to help any investigator who in, say, the 1810's had sought to study Cavendish's residual bubble.

Yet it seems that this was one of the unduly delayed advances in science. In a later chapter I shall make the point that very often new ideas or new experimental advances are recognized only if "the time is ripe." In a sense this was true of Cavendish's statement about the $\frac{1}{120}$ part of the nitrogen that was unabsorbed in his experiments. But on the whole one sees no reason why any time after the chemical revolution and particularly after the acceptance of the atomic theory (1860) argon might not have been discovered; though in this case the discovery of the other rare gases might have followed much later and the debate as to whether argon was an element or a compound might have been protracted.

Two footnotes to this case may serve to close the chapter. An American chemist, W. F. Hillebrand, had a sample of argon mixed with helium (another rare gas) before 1890 and failed to recognize his find. He had discovered that certain minerals when treated with acid gave off a gas which he reported as nitrogen. Ramsay's attention was drawn to this paper; he repeated Hillebrand's work and found the gas was not nitrogen but a mixture of argon and helium. The latter was recognized by its characteristic spectrum in an electric discharge tube as being identical with an element hitherto unknown on the earth but characteristic of the spectrum of the sun. Why Hillebrand did not investigate more thoroughly the gas he had remains an interesting question. Writing of his failure, after Ramsay's work, he said, "The circumstances and conditions under which my work was done were unfavourable; the chemical investigation had consumed a vast amount of time, and I felt strong scruples about taking more from regular routine work. I was a novice at spectroscopic work of this kind . . . It doubtless has appeared incomprehensible to you, in view of the bright argon and other lines noticed by you in the gas from clevite, that they should have escaped my observation. *They did not.* Both Dr. Hallock and I observed numerous bright lines on one or two occasions, some of which could be accounted for by known elements—as mercury, or sulphur from sulphuric acid; but there were others that I could not identify with any mapped lines. The well-known variability in the spectra of some substances under varying conditions of degree of evacuation of the tube led me to ascribe similar causes for these anomalous appearances, and to reject the suggestion made by one of us in a doubtfully serious spirit, that a new element might be in question."

One may speculate that for a conscientious government chemist intent on serving the practical needs of the time the question of pursuing a scientific lead seemed relatively unimportant. If this be so, this phase of the rare gas story may serve as a reminder of the effect of the climate of opinion on whether or not there are "prepared minds" (to use Pasteur's phrase) ready to follow up what chance offers.

Another footnote concerns a possible faulty generalization that can be drawn from Rayleigh's meticulous quantitative experimentation. For the reader this may be erecting another straw man, but since the fallacy in question is fairly widespread the issue is worth underlining. Indeed, Ramsay's biographer perpetuates what I venture to consider an erroneous view when he begins the story of the discovery of argon with a quotation from Lord Kelvin as follows:

"Accurate and minute measurements seem to the non-scientific imagination a less lofty and dignified work than looking for something new. But nearly all the grandest discoveries of science have been but the rewards of accurate measurement and patient long-continued labors in the minute sifting of numerical results."

Now it is true that "the minute sifting of numerical results" did in Lord Rayleigh's hands lead directly to the discovery of argon. But that does not prove the validity of Lord Kelvin's pronouncement. It merely proves that an accidental observation *followed up* may have its origin in painstaking measurements. But to argue from this that

merely pushing the accuracy of a measurement to another
decimal point is going to be fruitful is obviously absurd.
At what point accumulating scientific results becomes a
mere hobby like stamp collecting is always debatable. Filling in the gaps in our systematic knowledge is generally
recognized as legitimate, nay, even honorable scientific
labor. To the extent that there are real gaps in a systematic
scheme of classification there is undoubtedly need for careful observations. But one could endlessly refine the measurement of such physical properties of elements and compounds as density, electrical conductivity, solubility in
water, and so on. If such work is undertaken for practical
ends, well and good; or if it is undertaken to test some idea,
then it represents a fair scientific gamble. To be sure, "accurate measurement and patient long-continued labors in
the minute sifting of numerical results" may be undertaken
by an investigator just for the satisfaction that such work
yields. But in that event he must expect no more support
from the public than would a stamp collector.

All of which is not to disparage quantitative measurements. Without accurate measurements physics and chemistry would never have developed. But the significance of
these measurements lies in their relation to new concepts
and conceptual schemes, and above all in the way the numerical results may be manipulated by logical processes.
In the next chapter we shall consider some simple examples
to show the fundamental role in science played by measuring instruments and the mathematical formulation of
quantitative results.

CHAPTER SIX

Geometrical Reasoning and
Quantitative Experimentation

ONCE AGAIN I must ask the reader to return with me to the seventeenth century and the study of pneumatics. Torricelli's scheme, it will be recalled, yielded certain deductions that could be tested by experiment. The results obtained verified the deductions; people became increasingly convinced of the validity of the concept of a sea of air. These tests, however, were essentially of a qualitative nature. That is to say, accurate measurements were not required nor was there any mathematical manipulation of the numerical values obtained. The qualitative nature of this development is one of the features that makes the case history of pneumatics simple of comprehension, since most readers who are not scientists will close a book with horror at the first indication that algebra is about to rear its ugly head. Yet to rest an examination of scientific methods at this point would be to distort the picture grossly.

To be sure, qualitative experiments have been more than once of prime importance in the advance of the physical sciences. In biology until very recent times this has been the only type of experimentation. Therefore a considerable understanding of experimental science can be obtained by a study of cases where neither precise measurements nor complicated mathematical ideas are involved. Yet it is not an exaggeration to say that the sciences of astronomy, physics, and chemistry are built on a foundation of careful measurements made with ingenious instruments. Furthermore, the significance of these measurements rests on their relation to mathematical concepts whose manipulation has involved new inventions in the realm of abstract thought. Any understanding of science must therefore include some appreciation of the importance of measuring instruments and their improvement; likewise, there must be some notion of the way in which mathematical ideas are related to the observations in the laboratory. This chapter is consequently devoted to a consideration of *quantitative experimentation* and the use of mathematics. If the reader finds the going difficult after a few pages, he will be well advised to skip to the next chapter. He will then be once more in an essentially qualitative realm, but he should be well aware of the significance of what has been omitted.

No apologies are necessary for the elementary nature of the ideas and the mathematics which are presented in the following pages, yet no reader should be misled into thinking that the examples chosen are representative of seventeenth-century science. One need hardly be reminded that the seventeenth century was that of Galileo and of Newton, a century which started with the former's study of falling bodies and ended with Newtonian mechanics and the invention of the calculus. If one wishes to appreciate to the full the role of mathematics in the early development of theoretical physics, the work of these two masters must be studied. But the ideas involved are far too difficult to be handled in this volume, which aims at providing an elementary exposition of scientific methods. A consideration of the problem of moving bodies (kinetics and dynamics) may well confuse the layman rather than inform him as to the role of measurement and mathematics in the advance of science. Diligent study, including the solving of numerical problems, is essential if one is to grasp even the rudiments of the branches of physics commonly associated with the name of Newton.

Unless one wishes to lose all sense of intellectual history, however, it is necessary to place the experiments in pneumatics against the background of the change from the Aristotelian world view of the Middle Ages to the Newtonian picture of the eighteenth century. (Herbert Butterfield's *The Origins of Modern Science* is to be recommended in this connection). Pneumatics, it must be remembered, developed in the period between Galileo and Newton and at a time when far more complicated mathematical ideas than any we shall consider here were in process of formulation. The Aristotelian world view was still present but was changing rapidly. Astronomers were at work fitting into the heliocentric hypothesis of Copernicus the vast amount of numerical data obtained by careful observations made in this and the preceding century. Improvements in measuring instruments had long been a concern of astronomers; mathematics as applied to physical phenomena had been shown to be fruitful, particularly by Galileo. The logic and mathematics of the Middle Ages, the stream of deductive reasoning, were rapidly merging with the art of experimentation. Quantitative experimentation in many areas was proving the power that lay in this new combination.

Geometrical or deductive reasoning as applied to the physical world can be illustrated by the history of the branch of mechanics known as hydrostatics. Since this subject is so closely related to pneumatics, it provides a convenient introduction to a consideration of quantitative experimentation. The behavior of water in pipes and cisterns must have been the subject of observation and discussion early in the dawn of civilization. For our purposes we need probe no further back into antiquity than to Archimedes, who lived in the third century B.C. The story of his death at the hands of a Roman soldier at the fall of Syracuse is so well known as to serve as an identification of this early scientist. Almost equally well known is the story of his inventing a method of determining whether or not a golden crown was made of pure gold by weighing it first in air and then in water. The word "eureka" and the mention of a bathtub will recall at least some dim remembrance of a famous legend. The principle which Archimedes is believed to have first promulgated in connection with his very practical concern with assaying precious metals by physical means has long carried his name; it was, however, only one of a whole series of interconnected ideas about liquids and the pressure exerted by liquids which were transmitted to posterity by his writings. These became known to the western world in the sixteenth century and formed the basis for further discussion of the behavior of liquids when at rest, a subject known as hydrostatics.

The principles of hydrostatics are of significance because their development illustrates the way in which one line of inquiry in science has progressed from the earliest times to the present. The sixteenth- and seventeenth-century treatises on hydrostatics which amplified and expounded the principles of Archimedes read like textbooks of geometry. In the writings on this subject of Stevin of Bruges (published about 1600) and those of Pascal (written about 1650, published in 1663), for example, we find little or no reference to actual experimentation. Deductive reasoning from postulates enabled these early theoretical physicists to carry forward the ideas of Archimedes. In so doing they were following the pattern of logical thought established centuries earlier by Euclid. Rigorous analysis and careful reasoning rather than ingenious experimentation were the essence of their method.

Indeed, even today some logicians argue that it would have been a great waste of time for Archimedes (or anyone since) to have attempted to establish by experiment the basic principle of his method of assaying gold. The principles of hydrostatics like the principles of plane geometry, one may say, follow as a matter of logic from certain postulates. Or at least so Stevin, Pascal, and even some mid-twentieth-century writers claim. Whether this claim is completely valid is not an easy question to answer, and I shall postpone its consideration for a few pages until it is more evident just where the problem lies.

The important point is this: In the development of physics in the sixteenth and seventeenth centuries, the geometrical mode of reasoning was carried over into discussions of physical phenomena. This type of deductive reasoning placed relatively little emphasis on actual experimentation and a great deal of emphasis on demonstrations which *might* eventually be realized in practice but rarely if ever are. Indeed, from Pascal's treatise on hydrostatics and pneumatics it is impossible to tell which of the experiments described were ever actually performed.

Science and Common Sense

The contrast between Pascal and Boyle is striking. Boyle is the artful experimenter, the meticulous observer, the almost tiresome recorder of details; he did as much as anyone to establish the tradition of scientific experimentation. His intellectual forebears were the artisans who had successfully performed experiments over the generations to improve such arts as making metals. Pascal is the mathematician and logician, a spiritual descendant of the Greek mathematicians; he recognizes the need for an occasional real experiment to check a vital point in the argument, so to speak (hence the Puy-de-Dôme expedition); but the argumentation is in terms of logic and possible experiments, the actual observations play little or no part. Possible experiments—paper experiments, some people would call them—are employed throughout the exposition. Pascal in his writings represents one continuing strand which has gone to make up the fabric of modern physics. He is one of the forerunners of the theoretical physicist; Boyle is perhaps the progenitor of all experimentalists. In the theoretical tradition in more recent times we need mention only the names of Maxwell and of Einstein; in the experimental, Faraday and Lord Rutherford. Some of the great figures in science have combined both traditions as did Galileo and Newton.

Over the centuries the experimentalist and the theoretician have cooperated and the work of one has supplemented the other. At times traces of annoyance with each other have been manifested, however, and one of the earliest examples is found in Boyle's comments on certain of Pascal's alleged experiments. The French mathematician, said Boyle, does not state that he actually tried the experiments, "he might possibly have set them down as things that *must* happen upon a just confidence that he was not mistaken in his ratiocinations." He further pokes fun at Pascal for not giving sufficient details so that anyone could repeat the experiments if in fact they were ever performed. As an example of some of the things Pascal described that strained one's credulity, Boyle refers to an experiment in which a man is supposed to sit 20 feet under water and place a tube extending above the surface against his thigh. But, says Boyle, Pascal doesn't tell us "how a man shall be able to continue under water in a great cistern full of water 20 feet deep."

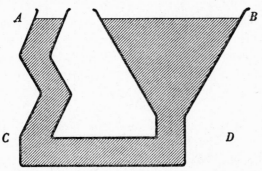

F<small>IG</small>. 17. Cross-sectional view of two vessels connected together by a tube. If water is poured into one vessel, the water level quickly becomes the same in both vessels.

THE PRINCIPLES OF HYDROSTATICS: TRUTH BY DEFINITION

With this general account in mind of the origin of the theoretical tradition in physical science, we can examine some specific problems in hydrostatics. First, we must recall a phenomenon which is sometimes summed up by the phrase that "water seeks its own level." The diagram (Fig. 17) will remind the reader that if two vessels are connected

and water is poured into one, the level will soon be the
same in both, *irrespective* of the shape. Clearly two *heights*
of water balance each other (*AC* and *BD*) although the
total amount of water in the two sides is very different. If
the water is poured in quickly there will be at the first some
movement of the two levels up and down, but this oscilla-
tion soon subsides and we say "equilibrium has been

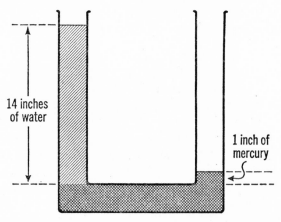

Fig. 18. Cross-sectional view of two vertical tubes, one containing
mercury and the other containing water, connected at their bottoms.

reached." We may note in passing that the concept of
equilibrium is of great importance in science. The principles of hydrostatics apply to equilibrium conditions, that
is, in situations similar to that illustrated by this example.
At equilibrium we find that a column of water about 14
inches high will balance a column of mercury 1 inch high
(Fig. 18); this seems reasonable since mercury is, volume
for volume, about 14 times as heavy as water.

The concept of pressure has proven convenient in for-

mulating the observations of such phenomena as those described in the last paragraph. This concept certainly has its
origins in everyday experience. The thrust or push of a
liquid is evident if one tries to prevent its flow out of an
opening in the bottom of a vessel by inserting a cork into
the hole or pressing one's hand over it. This thrust is due
to the weight of the liquid in the vessel, and it can be
shown that it depends on the depth of the liquid, its
density, and the size of the hole. If we make two holes, of
different sizes, in the bottom of a vessel containing liquid,

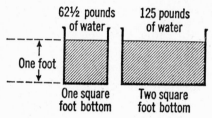

Fɪɢ. 19. Cross-sectional view of two vessels containing water. Although the total force on the bottom is different for each vessel, the
pressure at the bottom is the same in both vessels.

the thrust or force will be greater for the larger hole. But
the forces divided by the corresponding areas will be the
same. This force per unit area is called the *pressure* and
depends only on the density of the liquid and how far the
hole in the vessel is beneath the surface. Thus, while the
thrust of the liquid coming out of a hole depends on the
size of the hole, the pressure everywhere over the bottom
of the vessel is the same.

Suppose that we have two vessels with bottoms of area
1 and 2 square feet respectively, each filled with water
to a depth of 1 foot (Fig. 19). Since we know that water
weighs about 62½ pounds per cubic foot, the first vessel

must contain 62½ pounds, the second 125 pounds. For such vessels, with vertical sides, the force pushing down on the bottom of each vessel is equal to the weight of water in it. But the *pressure* on the bottom is the force divided by the area, which is 62½ pounds per square foot for both cases. An important generalization is that the pressure produced by 1 foot of water is always 62½ pounds per square foot, no matter how much water is involved or what shape of vessel. Thus a pressure can be specified by giving a depth of water: a pressure of 34 feet of water is the same as 34 x 62½ or 2,125 pounds per square foot.

We could use another liquid just as well: since mercury weighs about 14 times as much, volume for volume, as does water, a height of ¼₄ of 34 feet, or about 30 inches of mercury, corresponds to 34 feet of water. The two heights of water in Fig. 17 can be thought of as balancing each other because the pressure at the bottom of each vessel is the same. It is then not surprising that a column of water should be balanced by a column of mercury only ¼₄ as high (Fig 18).

Alternative methods of measuring liquid pressure are illustrated in Fig. 20, where an apparatus is shown in cross section. *A* and *B* are tight-fitting pistons, well lubricated so as to move easily, and *C* and *D* are tubes open to the air, the former containing water, and the latter liquid mercury. When the whole system is in equilibrium a weight of 100 pounds on the piston of an area of 100 square inches will balance a weight of 10 pounds on the piston of an area of 10 square inches. (If the fact that a weight of 100 pounds can thus be balanced against one of 10 pounds surprises the reader, he or she may take comfort in the knowledge that this has long been known as the hydrostatic paradox.) In both cases the pressure is 1 pound per square inch. And, as shown in the diagram, a column of water about 28 inches high or a column of mercury about 2 inches high will produce this same pressure. Pounds per square inch, inches of water or inches of mercury are all measures of hydrostatic pressure.

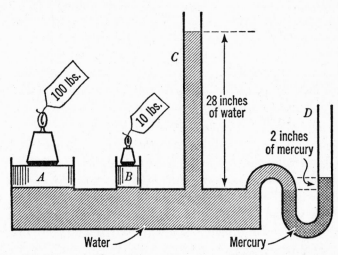

Fig. 20. Diagram illustrating various ways of designating pressure. The piston *A* has an area of 100 square inches and the piston *B* has an area of 10 square inches, so that in each case the pressure is 1 pound per square inch. This is equivalent to about 28 inches of water and to about 2 inches of mercury.

In the discussion of pneumatics we assumed that the reader had a general idea about liquid pressure and the balancing of columns of liquids. It is clear that unless the principles of hydrostatics had been well developed, Torricelli's notion of a sea of air would have been largely meaningless. In effect, what he did was to carry over to another fluid, air, the idea of balancing one column of a liquid against another; for just as water pressure may be visualized in terms of the height of a column of water, so air pressure may be thought of in terms of the height of a column of air.

What concerns us at this point, however, is the fact that hydrostatics as a branch of mechanics dealing with systems in equilibrium did not develop as an experimental science. Observed phenomena there were; but these were treated either as checks on the theory (like Pascal's use of the Puy-de-Dôme experiment) or as basic common-sense data. The type of reasoning employed may be illustrated by referring to one recurring line of argument, namely the impossibility of perpetual motion; this was a favorite methodological device of Stevin. Thus, in demonstrating the first theorem in his *Fourth Book of Statics* Stevin argues essentially as follows: Any designated portion of water in a vessel "maintains whatever position is desired in water," as otherwise this water would be in perpetual movement which would be absurd. This theorem, once established by this type of argument, then was used as a basis for other propositions.

To illustrate the use of the postulate of the impossibility of perpetual motion, let us examine a more modern application within the field of hydrostatics. Let us see how the use of this type of argument can establish the principle that the pressure below the surface of a liquid is the same in all directions. Imagine a point below the surface of a liquid at A (Fig. 21); imagine further two fine tubes inserted as shown, the one horizontal, the other vertical, pointed upward, both full of the liquid in question and joined in the manner shown in Fig. 21. If the pressure to the right is greater or less than the pressure down, then the liquid will flow around through the tube. But this would be perpetual motion which by postulation is impossible, hence the pressure down and to the right must be equal. (The same argument can be applied however the tubes are pointed.)

The example given, while oversimplified and cast in modern language, hardly distorts the use of deductive reasoning as applied by the earlier explorers of problems in mechanics. By the use of similar methods of reasoning one can establish Archimedes' principle "that a solid weighed

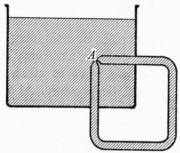

Fig. 21. Cross-sectional view of a vessel containing a liquid. A tube runs vertically from the point A down through the bottom of the vessel, around and back through the side.

in a liquid loses in weight the weight of the same volume of the liquid." It is important to realize that all these arguments apply to what may be called ideal liquids, that is, liquids under conditions in which certain assumptions necessary for the argument would hold true. Clearly one of these is rapid response by movement to alterations in pressure; another is uniform weight per unit volume (density) throughout. Let us consider the first assumption first; if into one side of the two-armed apparatus of Fig. 17 one pours sand instead of water it is obvious that one would have unequal heights in the two arms. The sand would "stick" we would say; it wouldn't come to "equilibrium" perhaps in a lifetime! Cold molasses likewise could yield very anomalous results over considerable lengths of time. In neither case would the arrangement respond to the requisite test for a system in hydrostatic equilibrium: a slight change in pressure on either area (adding more sand or more molasses) would not produce a rapid change. Water, alcohol, mercury, salt solutions, on the other hand, all meet the required condition of rapid movement under differences of pressure.

The significance of the second assumption (uniform density) can be made clear by imagining a deep well of pure water into which we lower a mercury barometer (Fig. 22). This barometer will serve us as an instrument for measuring the hydrostatic pressure. At the surface the pressure of the atmosphere alone holds up the mercury column (let us assume it is 30 inches); as we lower the apparatus, there is in addition the hydrostatic pressure. At a distance close to 34 feet below the surface the pressure will be twice the atmospheric pressure (remember, the water barometer has a column of water about 34 feet high). At this point therefore the mercury would read about 60 inches; 68 feet down the mercury will have risen to about 90 inches (if the barometer tube is long enough). I have purposely used approximate figures. For more precise numbers one need only know the relative weights of the same volumes of water and mercury and the relative densities, *provided* that the water in the well is of uniform density throughout. This proviso is met approximately if we have uniform temperature. Otherwise, as in the ocean, we would have varying strata of liquids with slightly different densities, for the density of water depends on the temperature. If we are going to push our analysis still further, we find

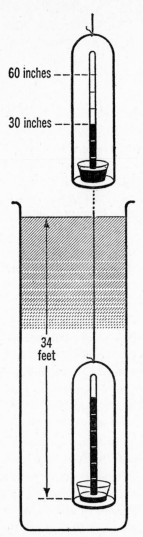

60 inches

30 inches

34 feet

FIG. 22. Diagram illustrating increase in pressure on descent into water. If a barometer is lowered into a well, the barometer will rise about 30 inches for each 34 feet of descent.

that even if the temperature were constant throughout a deep well there is a further complication. The density of the water increases slightly with the depth, for water while almost incompressible as compared with air is still slightly compressible.

Thus we see that an apparently self-evident proposition in hydrostatics turns out to be true only for a liquid that does not in fact exist. To be sure, as a close approximation, we may equate the behavior of water at constant temperature with the imaginary liquid; we may say that obviously the pressure below the surface of a body of water (at constant temperature) expressed in inches of water is equal to the distance below the surface also expressed in inches. Yet note that what appears to be almost a self-evident proposition or a definition of hydrostatic pressure is more nearly a definition of an ideal liquid, that is to say, a liquid whose density is independent of the hydrostatic pressure!

The principles of hydrostatics which can be established by the geometrical method of reasoning are thus seen to be principles about a fluid which is defined by the postulates employed. In actual practice the behavior of many liquids at constant temperature approaches closely that postulated for an ideal liquid. For most purposes the compressibility of water or aqueous salt solutions (sea water) can be neglected; so too can the small changes in density of water and similar fluids with changing temperature. Data obtained in the last century enable one to calculate how much error would be introduced by assuming uniform density throughout a depth of say 10,000 feet; it turns out to be small—less than 2 per cent. The sixteenth- and seventeenth-century writers were not far wrong *for practical purposes* in blissfully ignoring some of the realities exposed by later careful experimentation.

Let us return to our imaginary experiment with the barometer in the well or, better, a barometer in a deep lake. We translate pressure measured in inches of mercury into pressure expressed as feet of fresh water by multiplying by the number of inches in a foot and by the relative density of mercury and water at the temperature in question. This we have already done when we noted earlier that 34 feet of depth should increase the mercury column about 30 inches. Clearly one could measure depth below the surface with considerable accuracy by such observations of the hydrostatic pressure (with suitable corrections for changes of density of the liquid). Let it be noted, however, that the fundamental relation that applies to our ideal liquid of constant density throughout does *not* rest on the results of careful measurements.

It would be a pure waste of time to attempt to establish by experiment how closely the depth in feet corresponded to the calculated pressure expressed in feet of water. Anyone who set out to do this by lowering a barometer into a well, for example, would in effect be using a most indirect way of measuring the change in density of water with increased pressure. We feel quite certain from other ways of measuring the compressibility of water that elaborate experiments with very delicate instruments designed to record the hydrostatic pressure to, say, a thousandth of an inch of mercury would be required. It could be done but would throw no light on the principles of hydrostatics from which we deduce the approximate relation between the height of the mercury column in the pressure gauge and the depth below the surface. The *variables* in the practical case—the alterations in density of water and mercury with changes in temperature and pressure—can be measured with far greater ease by direct methods.

Are we then reduced to the statement that the principles of hydrostatics have no experimental basis, that they are the consequence of logical manipulation of arbitrary postulates? Obviously not. The approximate measurements inherent in even such qualitative observations that water seeks its own level are surely the basic data. Materials that do not come to equilibrium within a reasonable time when placed in a two-armed vessel such as is shown in Fig. 17 are not classified as liquids. For those that do a set of theorems can be constructed which can be checked by experiments. If the measurements are made with great precision, discrepancies will be found which can be correlated with other theorems about liquids (e.g., liquids change in density with change in temperature). In developing the principles of hydrostatics we neglect all factors other than those comprehended in our postulates about the ideal liquid. For example, referring to Fig. 20, we neglect the friction of the pistons; in the long tube containing the water column we neglect the attraction of the side of the tube for the water (capillary attraction, which with very narrow tubes is quite appreciable).

In short, by the use of imaginary experiments and logical arguments we construct a set of principles and from them draw deductions all of which correspond approximately to the behavior of real liquids. In so doing, those who first developed this branch of mechanics were the forerunners of the theoretical physicists of today. They were arguing, as did the geometricians, but were applying the modes of thinking of mathematicians to phenomena that were of increasing concern to experimentalists. That such procedures provided enormously powerful intellectual tools is evident even from the simple example given.

As the problems in physics became more complicated and involved, new types of mathematics had to be invented. As science progressed, the basic data were not provided by common experience but were obtained as the results of quantitative experimentation. This type of investigation involved the construction of precise instruments with which measurements of great accuracy could be made. From the eighteenth century on, a recurring type of experimentation is based on the desire of some investigator to increase the accuracy of certain quantitative data.

The desire for greater precision is in some scientists the equivalent of an aesthetic emotion. A frank appraisal of the vast amount of labor which has gone into quantitative experimentation aimed only at increased accuracy would show much wasted effort but occasionally great rewards. The famous Michelson-Morley experiment which provided the starting point for the theory of relativity is an example of what may result. The invention of a certain instrument and the development of elaborate procedures enabled scientists in the late nineteenth century to measure the speed of light with great precision. It was then possible to determine whether or not this speed was independent of the orientation of the apparatus with respect to the movement of the earth's surface relative to the fixed stars. The results obtained led, in the hands of Einstein, to revolutionary ideas. With these highly difficult matters, however, I cannot attempt to deal. Rather let me jump to the opposite extreme; to illustrate how new concepts arise from quantitative experimentation, let me revert to seventeenth-century pneumatics and discuss the discovery of Boyle's Law.

BOYLE'S LAW

Air is a highly compressible fluid; this fact clearly complicates the application of the principles of hydrostatics to pneumatics. For example, if we concentrate attention on the barometer, we can think of the column of mercury about 30 inches high (at sea level) balancing a column of air that reaches way up into the sky. But how high is the column? If the compressibility of air could be neglected, we would have only to know the relative weights of equal volumes of air and mercury at a given temperature and then do the necessary arithmetic. But a moment's consideration shows that this simple solution will not suffice. For as one goes up in the air the air "gets thinner"; in other words, the column of air constantly diminishes in density. How does it diminish? What is the relation between pressure and density? Or, if you concentrate your attention on a given weight of air, what is the relation between the volume of this air and the pressure? Neither Torricelli nor Pascal seem to have attempted to answer this question except in a qualitative way by referring to the analogy between a mass of sheep's wool and air. It remained for Boyle to provide the data and for Boyle's acquaintances to suggest the basic relationship. To trace this story we must return for a moment to Boyle's pump and his evacuation of the space above the barometer (Fig. 8).

It will be recalled that Boyle's experiment with the barometer and his pump (p. 81) involved observations in which the accuracy of the measurements was of little consequence. As the pump was operated the column fell; when the air was admitted the mercury rose; this is a typical qualitative observation. Boyle would have liked to relate the number of strokes of his piston to the numerical decrease in the height of the mercury column on each evacuation, but he was unsuccessful. The best he could do was to show that the smaller the vessel being evacuated, the greater the fall of the pressure on each stroke (the size of the pump cylinder was the same in all cases). In his first report we see him fumbling badly for a way of handling this problem. What he was concerned with was really some way of formulating the elasticity of air ("spring of the air") which would permit the use of mathematical reasoning. Boyle was an experimenter and not a mathematician. It is not surprising, therefore, that apparently the first hint of a fruitful idea came from one or two friends. They suggested the hypothesis that the force of the spring of the air was simply related to the volume; if you doubled the volume you decreased the "spring" by a half or, conversely, if you halved the volume (by a compression) you doubled the "spring." This was a generalization, deductions from which could be tested by experiment *provided* you had methods of measuring the volume and the "spring." Boyle's combination of a pump, receiver, and barometer was clearly a poor device if for no other reason than because of the leaks. A very simple method, however, soon forced itself on the experimenter's attention. I refer to the J-tube still used in every elementary physics laboratory to demonstrate Boyle's Law (Fig. 23). This tube was designed by Boyle, as we have already seen in a previous chapter, to confound the supporters of the "funiculus" theory.

Starting with the level of the mercury the same in both the long and short leg of the tube, Boyle poured in mercury and from time to time measured two lengths; one the height of the mercury in the longer leg above the level in the shorter leg (Fig. 23) and the other the distance between the closed end of the short leg and the level of mercury in that leg. The latter is a measure of the volume of the air in the short leg *provided* the tube is of uniform bore, and the former is the additional pressure on this air due to the experimental manipulation. The total pressure is then that of the atmosphere *plus this additional pressure.*

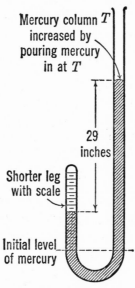

Mercury column T
increased by
pouring mercury
in at T

29
inches

Shorter leg
with scale

Initial level
of mercury

FIG. 23. Diagram of apparatus used by Boyle to collect the data on relation of pressure and volume.

In terms of inches of mercury, which we may remember is a convenient way of recording pressure, the total pressure on the enclosed air in the short leg is easily calculated by adding to the barometric reading the observed length in inches of the mercury column in the experiment. Boyle followed exactly this procedure. When he did so, he found that doubling the total pressure just about halved the volume; increasing the pressure fourfold decreased the volume to very nearly one-fourth. In general the hypothesis that there was a simple numerical relation between pressure and volume was confirmed. The "spring of the air" was greater the more one compressed the air, and the relation was one of simple proportion.

In more formal terms one may say that the *volume* of air in Boyle's experiment is *inversely* proportional to the pressure. If we let P_1 represent the initial pressure, P_2 a second pressure, and V_1 and V_2 the corresponding volumes, we may express the inverse proportionality by writing:

$$\frac{P_1}{P_2} = \frac{V_2}{V_1}, \text{ or } P_1V_1 = P_2V_2.$$

The substitution of actual numbers in an algebraic relationship usually facilitates one's comprehension. Let us assume the original pressure P_1 is 30 inches of mercury (approximately atmospheric pressure at sea level) and the original volume is 10 cubic inches. Then P_1V_1 is 30×10 or 300; if we now increase the pressure to 60 inches of mercury (P_2), clearly V_2 must be reduced to 5 cubic inches for 5×60 also equals 300. For the particular sample of air taken, 10 cubic inches, with the pressure expressed in inches of mercury, all products of pressure and volume (PV) must be 300 *if* the proportionality holds. In other words, we may say that as a first approximation, at least, the product PV is a constant. This is a common way of expressing what is known as Boyle's Law.

In Boyle's time everyone knew that heating a volume of air causes it to expand and cooling it causes it to contract. Therefore Boyle was aware that another *variable* which determines the volume occupied by a given weight of air is the temperature. He made a few very rough experiments to show that even highly compressed samples of air expanded on heating and contracted on cooling, but neither he nor his friends attempted to measure the relation of temperature change to volume change. Those experiments had to wait until the thermometer was made a more precise instrument. Of this development I shall have more to say shortly. But to anticipate, we may here note that experiments by eighteenth-century investigators showed that at room temperature a gas expands about $\frac{1}{530}$ of its volume for every Fahrenheit degree increase in temperature. This is a very appreciable effect. Therefore one always hastens to say that Boyle's Law is applicable to a body of gas only at *constant temperature*.

Even when the temperature is closely controlled, careful measurements of the relation of volume to pressure show that Boyle's Law is only an approximate formulation of the behavior of air or any other gas. The extent of the deviations from Boyle's Law depends on the nature of the gas, and in general these deviations are greater at pressures higher than atmospheric and less at lower pressures. Indeed, quantitative measurements made with great precision have shown that at pressures only a fraction of that of the atmosphere Boyle's Law becomes a very close approximation to the actual behavior of a gas. Just as we have spoken of an ideal fluid, so we may speak of an ideal gas. The first is defined in terms of the simple theorems of hydrostatics, the second by Boyle's Law. Indeed, while the phrase "ideal liquid" is rarely if ever used, the concept of an ideal gas has become commonplace among physicists and chemists in the last hundred years or so. An ideal gas is one for which the relationship $PV = a\ constant$ (at any definite temperature) holds true for all ranges of pressure. Many highly important principles connected with heat engines have been derived by considering imaginary experiments performed with such an ideal gas. The type of reasoning is analogous to that used by the early founders of hydrostatics; but it must be noted that the basic data were *not* provided by observations about commonplace phenomena such as liquids seeking their own level but by careful measurements of the type first made by Boyle. We see here a fine example of the blending of the two traditions, the geometrical mode of reasoning and experimentation; the latter, however, has now passed to the stage where *quantitative experiments* yield the essential data.

We can conclude the discussion of pneumatics and hydrostatics by what amounts to a footnote to a practical question I earlier raised and left unanswered. How high is the column of air that balances the 30 inches of mercury in a barometer at sea level? Let me remind the reader that what we need to know is the way the density of the air changes as we ascend away from the earth's surface. We shall have to proceed by first making an approximate estimate based on certain assumptions. *If* the temperature is constant (which it certainly is not) and *if* Boyle's Law holds for air (which it does approximately), we might expect to find a regular decrease in density as the distance from the earth increases. A few minutes' consideration indicates that the relation is not simple; there is going to be something tricky in the mathematics even with these simplifying assumptions, for as we go up in a balloon (or a plane) the pressure decreases (Pascal and Perier's Puy-de-Dôme experiment). As the pressure decreases a given weight of air occupies more and more volume (Boyle's Law). How does this work out numerically?

Boyle's Law could be expressed in terms of the relation

of pressure to the weight of a *given volume* of air. Thus at sea level mercury is roughly 10,000 times as heavy as air, volume for volume, while at a pressure of one-half an atmosphere (15 inches of mercury) the ratio is nearly 20,000, *provided* the temperature is constant and assuming that the deviations from Boyle's Law are negligibly small. Which of these values should we use in estimating the height of the column of air which balances the 30 inches of mercury in a barometer? Neither, of course. But for a small distance above the earth's surface the first figure will not be far wrong. Using it, one might predict, as a first approximation, that a rise of 120,000 inches (10,000 feet) from the surface of the earth would cause a drop in the barometer of $\frac{1}{10,000}$ as many inches, or 12 inches, so that the barometer would read 30 minus 12, or 18 inches, at an altitude of 10,000 feet. Even this approximation shows us that at 10,-000 feet the pressure is markedly reduced—nearly, in fact, to the value of 15 inches of mercury, or one-half an atmosphere, and the density of air has therefore dropped to nearly $\frac{1}{20,000}$ that of mercury. Actually, at a pressure of 18 inches of mercury the density would be $\frac{18}{30}$ times that at sea level, so that the density of air is only $\frac{1}{16,000}$ that of mercury. Using this figure we can compute that a rise of another 10,000 feet, to 20,000 feet, would cause the barometer to drop by only about 7.2 inches, since the air is here less dense than near to the ground, and the barometer at 20,000 feet would be then expected to read about 10.8 inches. Similarly, since the density of the air in the 20,000–30,000 foot region is even less, we can compute that a rise to 30,000 feet would drop the barometer by only about 4.3 inches more, to about 6.5 inches.

This would appear to go on indefinitely. For when we have risen so high that the barometer reads only 3 inches of mercury, the ratio of the density of mercury to that of air will have increased by a factor of 10 (since the pressure has dropped by 10) from that at sea level and will be 100,-000. It would take an additional height of 25,000 feet of air of this density to balance the 3 inches of mercury. But as we rise the density of the air continually decreases, so that when the pressure has fallen to 0.3 inches (and this must be purely an imaginary flight until one gets a rocket) the ratio of densities has risen to 1,000,000. And it would take an additional column of air 25,000 feet high and of this density to balance this 0.3 inches of mercury. If Boyle's Law can be applied there is literally no end to the atmosphere!

This method of estimating what would happen if one went straight up in a balloon in a sea of air at constant temperature is obviously a very crude approximation. Actually the density of the air would be decreasing regularly with every fraction of an inch of rise; there is no reason for recalculating the density every 10,000 feet. Theoretically we ought to recalculate it for each small increment, though we would commit a very small error in assuming a constant density of the air for every 10-foot interval, and if we should stop and take a new value for each 10 feet of ascent we would get a quite accurate picture of the situation. The smaller the interval, the more accurate the result. In general we note that as the pressure goes down the air gets thinner; therefore longer and longer distances upward correspond to a given decrease in pressure. Or, looking at it from the point of view of a balloon descending from one level to another, one may say that the vertical distance down which corresponds to a given increase in pressure depends on the initial pressure.

The contrast with the situation in the case of a liquid

is striking. If you move a pressure gauge up or down in a liquid, we have seen that you can equate change of pressure with distance (at least as a very close approximation). But with a balloon you have to know not only the change of pressure but the initial pressure before you can calculate the distance; what is true at 30,000 feet is not at all true at 10,000 feet. Anyone who has used a barometric altimeter for measuring heights in climbing mountains may have noticed the peculiar arrangement of the scale; the numbers are crowded together more closely at the high end then at the lower. As some readers will be well aware, the relation between height and atmospheric pressure is approximately logarithmic.

By the use of the calculus it is possible to show that whenever for small increments of one variable, say distance, another variable, say pressure, changes in direct proportion to the quantity itself (pressure), then a logarithmic relation is at hand. If we let the symbol $\triangle h$ represent a small change in height and $\triangle p$ the corresponding change in pressure, then to the degree that Boyle's Law expresses the behavior of air we can show that $\triangle p/p$ is directly proportional to $\triangle h$ where p is the atmospheric pressure at the point in question. The application of the calculus then enables us to convert this relationship, which holds only for very small changes, into one for *any* change; the mathematical manipulations turn out equations in logarithmic form; *the difference in the logarithms of the pressure at two different heights is proportional to the differences in the heights.*

This excursion into what *would* happen if the atmosphere were at a constant temperature is of significance as illustrating how mathematical formulations may be of significance in handling physical problems. This application of Boyle's Law is a very simple case of the use of the calculus to solve a problem which otherwise can be handled only by a series of rough approximations. Readers to whom the word logarithm carries no meaning need not worry about the mathematics. For it is only as an example of the use of mathematical manipulation of quantities that the preceding paragraphs have value. Actually, the variations in temperature in the atmosphere are so great that the calculations made by Boyle's Law even with the aid of the calculus are only approximate. Another variable is the humidity. Nevertheless, for relatively low elevations, if the atmospheric pressure at a given spot remains constant, the barometric reading will give fairly accurate estimates of height up to 10,000–15,000 feet (within 100 feet or so). The following comparison of recorded pressures and pressures computed by the method previously described may be of interest.

Height in Feet	Typical Barometer Readings (inches of mercury)	Pressure Calculated Using Simplifying Assumptions
0	30	30
10,000	21	18
20,000	14	11
30,000	9	7

Of course, it should be understood that just as the barometer reading varies within a range of an inch or so at sea level with changes in meteorological conditions, so the readings listed above are only typical.

The curious reader may well be wondering whether there is "in fact" a limit to the atmosphere. According to modern concepts, there is a zone a few hundred miles above the earth's surface where the force of gravity is not suf-

ficient to hold the molecules to the earth. There appears
to be here "in fact" an upper limit to the atmosphere of the
earth.

An understanding of the intricate relationship between
concepts, conceptual schemes, and experiments is the essence of understanding science. Some concepts have undoubtedly arisen as a result of qualitative experiments or
observations; many more, it is probably safe to say, have
arisen through quantitative experimentation. It is hard to
know whether to use the word "arisen"; perhaps "developed" would be better because so often vague ideas
have become scientific concepts as a consequence of experimental findings. The significance of satisfactory measuring instruments in this connection can hardly be overesti-
mated. One case from the eighteenth century may be cited
to drive this point home.

The history of two simple concepts which are described
in every high-school physics book, namely specific heat
and latent heat, is highly illuminating. These terms and the
ideas behind them can only be understood in relation to a
measuring instrument, the thermometer. Temperature is
for people of the civilized world today a common-sense
idea; it is this that makes relatively easy its manipulation
in physics textbooks or the use of such phrases as "constant temperature" in a book like this for the general
reader. Compressing a great deal of history into a few
lines we may say that the concept of temperature developed from crude common-sense ideas about one object
being hotter or colder than others. The ability of the human
organism to make the distinction between cold water and

hot water, for example, is certainly one basic factor in this whole story. But surely not the only one; for the effect of fire in making water boil, the action of a flame on all sorts of materials, as in making glass and metals, as well as the change in color (red heat) are other obvious manifestations of changes which correspond to something related to fire.

Though thermometers were known as early as the first third of the seventeenth century, they did not become satisfactory measuring instruments till nearly a hundred years later. The members of the Accademia del Cimento in Florence about 1650 had instruments closely resembling the simple household thermometer today *except* as to the scale. And the exception is all-important. From the middle to the end of the seventeenth century thermometers were built which contained alcohol or mercury in a bulb and a column in a sealed-off tube. There were various changes in design but the significant event occurred when a relatively simple scale was first adopted which could be easily reproduced in different laboratories. By the turn of the century it was beginning to be common to use *two fixed points* which corresponded to the temperatures of easily reproducible situations. The freezing point and boiling point of water taken as 0° and 100° yielded one scale, the Centigrade; another scale, the Fahrenheit, which came into use at the same time, placed the zero as the temperature of a certain mixture of ice and salt and the boiling point of water at 212°.

Neither the differences in the scales nor the details of their origin need detain us. The important point is this: as soon as instrument makers were able to deliver to the scientific investigators of the day reliable instruments with a satisfactory scale, some interesting things began to happen. Before this time the thermometer reading was referred to a single point only. This would enable a person to record that

some object (or place) was hotter or cooler than another
but not how much. Now one could speak of differences in
temperature in terms of degrees Centigrade or Fahrenheit,
according to the scale employed. Once this was done, questions appeared that had hardly been visible before. People
now began to talk about heat in different terms from those
they had earlier employed. Such an occurrence in the history of science we see reflected time and time again: a new
measuring instrument or an improved one may be the
means of opening up a whole new field of inquiry.

The idea of a material substance related to what we now
call heat can be traced far back into antiquity. The common-
sense assumption would seem to be that when you stood
near a fire and became hot, some part of the fire passed into
your body. The Aristotelian picture of the universe, which
completely dominated European thought until well into
the sixteenth century, accommodated in its own way the
simple phenomena connected with ideas of "fire" and "hot"
and "cold." We shall not dwell on the way in which the concepts of the four elements of Aristotle, earth, air, fire, water,
could be manipulated to account for boiling, melting, freezing as well as combustion, though any careful analysis of
the concepts of temperature and heat would have to take
into account this highly important part of a long history.
It is the eighteenth-century portion of the story to which
I am directing attention in order to illustrate the way new
measuring instruments may play a determining role in
scientific thought.

The historical records are far from satisfactory, but it appears that experiments which led to the concepts of latent
heat and specific heat were carried on probably independently by a Scotsman, Joseph Black, and an Englishman,
Henry Cavendish. To Black quite properly goes the credit,

for he carried his ideas further and made them known through his lectures at the University of Glasgow. (They likewise received much attention because they provided to some degree a theoretical background for Watt's invention of the steam engine. Unfortunately, there is little evidence as to how far Watt stimulated his friend Black or vice versa.) Before Black and Cavendish, however, other eighteenth-century scientists were trying quantitative experiments with the new types of thermometers (i.e., those with a satisfactory scale). Such questions as these were asked: If equal weights of water at 40° F. (Fahrenheit) and 100° F. are mixed, what will be the temperature of the mixture? If such different liquids as mercury and water are similarly heated and mixed, what will be the result?

Rather than follow the arguments involved in experiments concerning the mixing of materials, I shall attempt to give a rough idea of the concept of specific heat by referring to another type of experiment which was employed by Black. This could be readily performed at home today with very little equipment. Take equal volumes of water and mercury and place them in thin glass vessels (two wineglasses, for example), put a thermometer in each, take them outdoors on a cold day, let them stand until both liquids have reached a temperature of, say, 50° F. (Move the thermometer around in both liquids to be sure of fairly uniform temperature throughout.) Take the two glasses into a warm room (say about 70° F.); place them side by side on a table, moving the thermometer in each, and note the time required for the thermometer to rise to 60° F. in each case. Then repeat the experiment and see how reproducible the results may be.

With good luck and patience an experimenter will find that within 10 per cent or so the *relative* times required for the rise of 10 degrees in the mercury and water are the same for all his experiments; and the ratio of these times is about two. Volume for volume, liquid mercury "warms up" a little more than twice as fast as water; if we express the results not in terms of volumes of liquids but in terms of weight, since mercury is 13.6 times as heavy as water, the difference is even more striking. Weight for weight, mercury "warms up" about 27 times (2×13.6) as fast as water. This ratio is found to hold irrespective of either the size of the samples of water and mercury employed in the experiments or the extent of the rise of temperature.

Up to this point nothing very interesting has developed. Quantitative observations like this could be made with increasing care and with a great variety of substances without leading to any advance in science. This is a point which I wish to emphasize even at the risk of straining the reader's patience unduly by belaboring such an oversimplified example. Measurements by themselves do not yield new concepts; those scientists who have made great advances with the aid of new or improved measuring equipment have known what to measure because they were able to bring in a new concept or conceptual scheme at just the right moment in the drama. In connection with some hitherto untilled fields of inquiry (particularly in connection with the study of man) I have heard an argument that runs essentially as follows: devise a measuring instrument, make a vast number of measurements with control of all the various variables, classify the results, and lo and behold, out will pop a new scientific principle! This is nonsense, a caricature of one type of phenomenon in the history of science. New measuring instruments *may* be significant and often have been significant in the hands of imaginative thinkers, but there is no guarantee of scientific advance down this or any other road.

But to return briefly to the difference in the speed of "warming up" of the mercury and the water. One way of interpreting the phenomena would be to think in terms of an invisible fluid with negligible weight which flowed at a definite rate from the warmer environment into the cooler liquids. If that assumption be valid, then the time noted for a certain rise in temperature (10 degrees) would be the measure of the amount of this fluid which had flowed into each vessel; this fluid we might call *heat*. Following along this line of thought we can speak of water and mercury having a different capacity for heat, for it takes about 27 times as much heat to raise water 10 degrees as it does to warm up the same weight of mercury 10 degrees. Black, arguing in part in some such fashion, arrived at the concept of *specific heat*. Specific heat can be defined as the heat capacity of a substance compared with that of water on a weight-for-weight basis. It is convenient to construct a scale of specific heats with reference to water taken as having the numerical value of one; in this case the specific heat of mercury is about $\frac{1}{27}$ or about 0.037 (accurate measurements yield the figure 0.033).

The concept of specific heats, it would seem, should enable one to formulate the temperature changes which occur when liquids of different temperature are mixed. For example, it is possible to calculate that if equal *volumes* of water and mercury at temperatures of 40° F. and 80° F. are mixed, the temperature of the mixture should be not 60° F. but about 52° F. if the specific heat of mercury is .033. This calculation is confirmed by experiment. Indeed, this type of experiment provided for years the most accurate method of determining the specific heats of various substances.

Black and his contemporaries were also able to make significant measurements in connection with the common phenomena of freezing and boiling and to formulate the results. Again a simple home experiment will serve to illustrate the development of a new concept. Put a given weight (say an ounce) of chopped ice in a wineglass, place the glass in a warm room (perhaps at 70° F.); stir the ice from time to time with a thermometer and note the time. You will find the thermometer registers very nearly 32° F. until almost the last of the ice is melted, which will take some time. What has been going on according to the ideas advanced in the preceding paragraph? Heat has been flowing in and melting the ice to water, we may say. Can we estimate how much? Yes, in *relative* terms, if we compare the time required to warm up a given weight of water under the same conditions. One way to proceed would be as follows: place 2 ounces of ice and 2 ounces of water previously cooled to 33–34° F. in a wineglass; in another glass place 4 ounces of water cooled to 33–34° F.; set the two glasses side by side in a room at 70–75° F. It will be found that approximately 1 ounce of ice will melt in the course of one hour (the volume of the water will increase to 3 ounces). In the second glass the temperature of the water will rise 8–10° F. in the first quarter of an hour; this rise is at the rate of 32–40° F. per hour and may be regarded as a measure of the amount of heat flowing in one hour into the water-ice mixture. If the idea of specific heats is satisfactory, 4 ounces of water warmed through 32–40° F. is the equivalent of 128–160 ounces of water warmed through 1° F., and this amount of heat appears to be required to melt 1 ounce of ice. Black called this heat the *latent heat of melting* and measured it in several ways, including one essentially the same as that just outlined. We often refer to it as the heat of fusion. If we adopt as a unit of heat the amount necessary to raise a certain weight of water a certain number of degrees, we can clearly express this latent heat in the units thus defined.

The concept of latent heat can also be applied to the phenomenon of boiling. It turns out that as much heat is absorbed in turning one pound of water into steam as is required to raise 970 pounds of water 1° F. The same amount of heat is given off when one pound of steam is condensed. (This can be determined by running steam into a mass of water and noting the rise in temperature for every pound increase in weight due to the condensed steam.)

In developing the ideas of specific and latent heats, I have used a conceptual scheme involving an assumption that heat is a subtle fluid of a certain sort. The idea of heat as a fluid proves, however, not to be an essential part of the argument, for as everyone is well aware, we now associate heat with the motion of small particles. Yet the notion that heat was "something" that could flow probably assisted in the rapid development of the concepts of specific and latent heat. By the middle of the nineteenth century, however, everyone had abandoned the idea that there was a "caloric fluid" which was present to a greater degree in hotter bodies than in colder ones or which was taken up by the particles of ice as they turned to water.

The story of the rise and fall of the caloric theory is a profitable one to study for many reasons. But a further consideration of this topic would require too much space. I have endeavored to center attention here not on the conceptual scheme but on two particular concepts which outlived the conceptual scheme. The fundamental definition of specific and latent heat used today is exactly as it was left by Joseph Black. By means of a measuring instrument an investigator with rare insight was enabled to formulate new concepts precisely from quantitative experiments. By very simple mathematics it was possible for him and his immediate followers to develop a method of calculating and thus predicting a vast number of thermal phenomena. Before the calibrated thermometer arrived on the scene there was no body of principles connected with temperature and heat; within a generation of the advent of this instrument the outlines of a highly satisfactory and extremely useful system had been evolved. No simple case seems better to show the relation of measuring instruments, quantitative experimentation, and ingenious reasoning than the work of Joseph Black on heat.

MATHEMATICAL TRUTHS AND
PROBABLE KNOWLEDGE

Experimental observations, however precise may be the instruments, always are subject to error. Some of these errors may be "corrected" by noting the variations in some measureable quantity (e.g., temperature) and making suitable calculations based on other measurements. This procedure has been cited in connection with the possible measurement of depth by noting changes in hydrostatic pressure. But over and above such types of error there are the uncertainties of the measurement; almost without exception the observer finds that repeated observations yield slightly different values; in recording his results he is apt to give a number followed by some such phrase as plus or minus a certain error. Anyone using a measuring tape to find a distance will have had the same experience and will have said, "The room is about 10 feet 10 inches long but I am not certain as to the fractions of an inch"; he could express the result by writing 10 feet 10 inches ± 0.5 inches, meaning the most probable value lay between 10 feet 9½ inches and 10 feet 10½ inches.

The question may be raised of course as to how reliable is our belief that any experiment can be repeated. This problem has already been referred to in an earlier chapter: our belief in the uniformity of nature seems to me to be rooted in common sense. Some modern philosophers, however, consider all knowledge derived from experience as only probable, whether formulated in common-sense terms or in most erudite scientific terminology. Those who argue in this fashion seem to me convincing; all knowledge of what we call the physical world is thus spread out over a spectrum of probabilities with one end so highly probable that in our usual frame of mind we consider it as "quite certain." The type of knowledge represented by mathematics, however, is now generally regarded as being in an entirely different category. Such knowledge is certain because it follows by logical processes from a series of definitions. Before the nineteenth century the philosopher's views about the truths of mathematics were quite otherwise. But the discovery of various types of geometries and the subsequent work of logicians have led many people to conclude that the truths of mathematics are not unlike those we assert when we say 12 inches = 1 foot.

The theorems of Euclid follow from the postulates; but other theorems follow from a different set of postulates. As a first approximation, Euclidean geometry is a "true" representation of what we actually find if we set up geometrical figures on a "flat" surface. The approximation is very close, indeed, but the "truths" of Euclid like those of other geometries (which do not represent figures on a flat surface) are quite independent of any actual measurements. When we consider vast distances and motions of particles with very high speeds, problems arise as to whether Euclidean geometry is adequate for the formulation of the results. But here we come once again to the boundary of the difficult area of relativity theory and quantum physics. For our present purposes it is sufficient to point out that mathematical systems deal with abstract ideas. The manipulation of these ideas through logical processes yields a vast array of powerful tools which can then be used in connection with the results that flow from observations and experiments.

The simple cases of hydrostatics and Boyle's Law have been dealt with in some detail in this chapter in order to indicate how abstract ideas can be of the greatest assistance to the experimentalist. The concepts of an ideal liquid and an ideal gas have proved of immense value to those who study the behavior of specific materials like water and air. The certain truths of mathematics yield formulations into which may be fitted by various devices the probable knowledge obtained by observation and experimentation. The concepts and conceptual schemes which comprise the fabric of modern science are found on careful analysis to be a strange mixture of mathematical truths and quantitative observation; the mathematician, the instrument maker, and the experimenter all have had a hand in the making of this strange fabric. For the balance of this book relatively little will be said about quantitative measurements and almost nothing directly about mathematics. Nevertheless it will be taken for granted that the reader knows that the conceptual schemes of modern physics and chemistry were made possible only by the development of new tools in mathematics and new instruments for measurement.

To hold the balance even between theory and practice in any book on understanding science is extremely difficult; even more so is the task of giving an appreciation of the importance of both qualitative and quantitative reasoning and experimentation. One can only emphasize that the history of modern science shows there is no one method, no one sure road to progress; at one period advance is rapid by one group of procedures, at another time by another. Mathematics and measurement are not to be unduly worshiped, nor can they be neglected by even the lay observer.

CHAPTER SEVEN

The Origin of a Conceptual Scheme: The Chemical Revolution

THERE IS no mathematics in this chapter: nonetheless, the reader who knows little or nothing about science may find it difficult. For what I shall attempt to outline is a scientific revolution of the first importance. It would not be too much to say that this case is concerned with the birth of chemistry, for Lavoisier, the central figure in the story, has long been called the father of modern chemistry.

It is not as an epoch-making event in scientific history, however, that I am including this somewhat complicated story. Rather I am proposing to use it to illustrate the steps by which a conceptual scheme matures as the result of experimental observations. For it so happens that we have a rather complete record of the way Lavoisier proceeded in developing his new ideas. We also can see clearly how difficult is the interpretation of experimental results, for just on the verge of the final step Lavoisier is diverted by a faulty interpretation of an experiment. We are here permitted to follow the course of a revolutionary scientific discovery in much the same fashion as a slow-motion picture of a critical moment enables us to understand what happened in a football game. There are other interesting aspects of the case as well. The fact that the time must be ripe for a scientific discovery to be effective is well illustrated. Likewise, the significance of controlling variables is apparent; for a chemist this often means controlling purity of materials. Finally we shall see how one conceptual scheme may be a block to the acceptance of another and how ad hoc assumptions can be used for a time to rescue a theory which is under fire.

The easiest way to understand the revolution in chemistry associated with the name of Lavoisier is first to describe the phenomenon of combustion in terms of modern concepts and then to show how for nearly a hundred years other ideas prevailed. Almost every high-school student of chemistry "knows" that air is primarily a mixture of oxygen gas and nitrogen gas and furthermore that when a candle or a match or a cigarette "burns" heat and light are being evolved by a chemical reaction involving oxygen. This reaction is called "combustion." If we burn enough material in a closed space the combustion stops because the oxygen is used up. What is it that is burning? Some but not all of the students will say that it is a group of carbon compounds, and some will add that the products of combustion are carbon dioxide (CO_2) and water (H_2O). If you heat molten tin in air at a high temperature for a long time, the bright metal becomes covered with a scum and this scum obviously is not a metal. What has happened? A combination with oxygen has occurred—an oxide was formed—the good students answer. Correct. Suppose we heat this nonmetallic substance, this oxide, with carbon. What would happen? The carbon would combine with the oxygen, giving an oxide of carbon and leaving the metal. This is what happens in making iron from iron ore, a sophisticated layman will tell you.

All very simple and plain. And you can set students to work in high-school laboratories to prove it. Yet at the time of the American Revolution not one philosopher or experimentalist out of a hundred could have given you an inkling of this explanation which we now designate as "correct." Instead, they would have talked learnedly of "phlogiston," a name perhaps unfamiliar to all but the chemists who read this book. Nearly a hundred years after Newton, and still scientists were baffled by such a simple matter as combustion! This fact needs to be brought home to all who would understand science and who talk glibly of "*the* scientific method."

The chemical revolution was practically contemporary with the American Revolution and thus just preceded the French Revolution. Lavoisier, the man who singlehanded, building on the work of others, made the chemical revolution, lost his head at the hands of the Revolutionary Tribune in 1794 (though he was by no means hostile to the basic aims of the great social and political upheaval). Whether or not he was betrayed by a scientific colleague (Fourcroy), who was at least an ardent supporter of the extreme party then in power, is an interesting historical question. Indeed, this case history abounds with matters of tangential historical interest. Another prominent figure in the final controversy was Priestley, a Unitarian clergyman, who was made an honorary citizen by the French Assembly and fled to America in the very year of Lavoisier's execution to escape a reactionary English mob. There is no lack of material to connect science with politics in the late eighteenth century, but for our present purposes all this is beside the point.

Before tracing the steps by which Lavoisier was led to his new conceptual scheme one must realize where he started. He and his contemporaries inherited a theory of combustion—the phlogiston theory. Indeed, the case at hand might well be called the overthrow of the phlogiston theory by the oxygen theory, for the new conceptual scheme which Lavoisier developed made the whole notion of phlogiston quite unnecessary. Yet his contemporaries were by no means quick to grasp this, and the last stand of the phlogiston theory (to be considered later in this chapter) is an interesting example of the tenacity with which old ideas persist.

THE SIGNIFICANCE OF THE PHLOGISTON THEORY

The phlogiston theory in its day was, we must first realize, a distinct step forward. In the sixteenth and seventeenth centuries those who were interested in making some sense out of what we now call chemistry were wandering in a bewildering forest. From the alchemists and the practical men, particularly the metal makers, they had acquired a mass of apparently unrelated facts and strange ideas about "elements." The earth, air, fire, and water concepts of Aristotle were still hovering over them. Boyle in *The Skeptical Chymist* had done something in the 1660's to clear a space in the tangled underbrush of fact and fancy so closely interwoven and cemented by strange words. Let us look at some of the common phenomena that had to be explained by Newton and his contemporaries at the end of the seventeenth century, that is to say, fitted into a conceptual scheme. Metals could be obtained by heating certain materials with charcoal (the ancient art of winning metals from their ores). Metals were at first sight very much the same; they had similar superficial properties. (Even today the classification of substances into metals and non-metals appeals to common sense.) Other solids were called "earths" (oxides for us today) and still others such as charcoal or sulfur were "combustible principles." Some earths when heated with charcoal yielded metals. This process could be reversed, for often but not always the metal (for example, tin) yielded, on heating, an earthlike substance. From such an artificial earthlike substance (an oxide in modern terms) the metal could be regained if the earth was heated with charcoal. A pure earth of this sort might be called a calx; the process of forming it by heating a metal was "calcination."

How were all these facts, inherited from the Middle Ages and before, to be fitted together? By the introduction of a principle called phlogiston, closely related to Aristotle's old element, fire, although the relationship was never clear. To those who sought for clarity it seemed evident that there must be some common principle involved in the process of making various metals from their calxes and vice versa. Let us call this something phlogiston, they declared in effect. When phlogiston was added to a calx you had a metal, when you removed it from a metal a calx was formed; phlogiston was in a sense a metalizing principle. Note there is a common-sense assumption more or less implied in this line of reasoning: except for gold, and occasionally a few other metals, it is the calxes, *not* the metals, that occur in nature. Therefore these calxes seemed to be the simpler materials; it appeared that something must be added to them to make them metals. Since metals were so alike, the "something" was obviously the same in all cases.

We shall call it phlogiston, said Becher and his pupil Stahl in a series of books published in the period 1703–31.

Here was a key to unlock a maze, and it was immediately accepted. Here was a concept that provided a pattern into which a mass of otherwise unrelated phenomena could be fitted. Substances were rich or poor in phlogiston: this seemed easy to establish. What was phlogiston itself? It probably was never to be seen. Substances rich in phlogiston easily took fire and, indeed, fire was perhaps a manifestation of phlogiston or worked with it at least. (To some, fire was still an element.) Charcoal was a phlogiston-rich material and on heating with a metallic calx gave up its phlogiston to the calx, making a metal. By itself charcoal burned, the phlogiston appearing as fire or combining with the air. Sulfur was found free in nature; it burned when heated and yielded an acid, vitriolic acid (sulfuric acid in modern terms). Clearly, this sulfur was only vitriolic acid highly "phlogisticated"; the burning set the phlogiston free and yielded the acid.

We can write these changes in diagrammatic form to illustrate how the chemists of the eighteenth century thought:

Calx + phlogiston (from charcoal) ⟶ metal

Metal heated in air ⟶ calx + phlogiston (to the air)

Charcoal burned ⟶ phlogiston (to the air) accompanied by fire

Phlogisticated vitriolic acid (sulfur to us) burned ⟶ phlogiston (to the air) + vitriolic acid (sulfuric acid).

The phlogiston theory was almost universally accepted at the time of the American Revolution and was the basis of the chemistry then taught to college students as part of their instruction in natural philosophy. The lecture notes of Professor Samuel Williams, the Hollis Professor of Mathematics and Natural Philosophy, 1780–88, at Harvard illustrate the convincing way in which the phlogiston theory can be presented to a class. "Take some combustible substance and let it be inflamed or set fire: In this state inclose it in a vessel containing a small quantity of atmos-

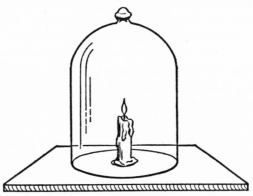

FIG. 24. When a bell jar is placed over a lighted candle, the candle will soon go out. According to our modern conceptual scheme the oxygen in the jar has been used up. According to the phlogiston theory the air has been saturated with phlogiston and hence can hold no more.

pherical air [Fig. 24]. *Effect:* The combustion will continue but a short time and then cease. Part of the combustible substance is reduced to ashes and the other part remains entire. And the air appears to be changed and altered. . . . Here then we have a representation of what the chemists call phlogiston and of the air's being loaded with it. In the confined air the combustible matter continues burning until the air becomes loaded with something that prevents any further combustion. And being confined by the closeness of the vessel, whatever the matter be with which the air is loaded, it is confined within the vessel and cannot escape. . . .

"It seems, therefore, from this experiment that phlogiston *must be a real substance,* and that the air is loaded or saturated with it. For what can the inclosing the combustible matter in the phial do but to prevent the escape or dispersion of some real substance? And is it not evident that so long as the air can receive this substance from the combustible matter so long the body will continue burning; and that as soon as the air is saturated and can receive no more of the phlogiston, the combustion must cease for no more phlogiston can escape or be thrown out from the burning body. And therefore when fresh air is admitted to receive phlogiston, the combustion will again take place . . . And hence are derived the phrases of *phlogisticated* and *dephlogisticated* air. By phlogisticated air is intended air which is charged or loaded with phlogiston, and by dephlogisticated air is meant air which is free from phlogiston; or which does not contain this principle element of inflammability." The italics are mine, but even without them Professor Williams is a convincing exponent of the phlogiston theory.

SCIENTIFIC DISCOVERIES MAY BE DISREGARDED

There was one very simple flaw in all this argument and the interesting thing is that this flaw was known 150 years before the phlogiston theory was even shaken, much less overthrown. This is a beautiful illustration of a principle in the strategy of science, namely that the time must be ripe for a scientific discovery to be regarded as significant. As early as 1630 (note the date—before Boyle was born) a Frenchman, Jean Rey, studied the calcination of tin and showed that the calx weighed more than the tin from which it was formed. More than that, he gave an explanation closely in accord with Lavoisier's ideas of 150 years later. For he said, "this increase in weight comes from the air, which in the vessel has been rendered denser, heavier, and in some measure adhesive . . . which air mixes with the calx, . . . and becomes attached to its most minute particles. . . ." Boyle confirmed the increase in weight of metals in calcination in 1673 but added no support to Rey's shrewd guess (it was little more) as to the reason. In fact, if anything, he led subsequent investigators astray. At least in retrospect, it seems that if he had followed up his own experiments only a little more boldly the phlogiston theory might never have been proposed or, if proposed, never accepted seriously. Yet it is all too easy to construct imaginary history. I doubt if even a still greater genius than Boyle could have discovered oxygen and revealed its role in combustion and calcination in the seventeenth century. Too much physics as well as chemistry lay under wraps which were only slowly removed by the labors of many men.

At all events, Boyle put forward the hypothesis that fire, the Aristotelian principle, had passed through the walls of the glass vessel used and combined with the metal, thereby giving it weight. This was, of course, not the same as the phlogiston theory formulated a generation later; in a sense it was the opposite because according to Boyle something was *added* to the metal in calcination, namely fire, while according to the phlogiston theory something, namely phlogiston, was *removed*. But Boyle's writings did focus attention on the heat and flame (a characteristic of fire and calcination) rather than on the air which had figured in Rey's explanation.

Rey's ideas about the air seem to have been lost in the subsequent 150 years, but the facts of calcination were not. That a calx weighed more than the metal was known throughout the eighteenth century, but this fact was not recognized as being fatal to the phlogiston theory. Here is an important point. Does it argue for the stupidity of the experimental philosophers of that day? Not at all; it merely demonstrates that in complex affairs of science one is concerned with trying to account for a variety of facts and with welding them into a conceptual scheme; one fact is not by itself sufficient to wreck the scheme. A conceptual scheme is never discarded merely because of a few stubborn facts with which it cannot be reconciled; a conceptual scheme is either modified or replaced by a better one, never abandoned with nothing left to take its place.

Not only was it known in 1770 that a calx weighed more than the metal from which it was formed (which to us means that something must have been taken up in its formation), but Boyle himself back in the 1660's showed that air was necessary for fire. John Mayow and Robert Hooke at about the same date had written about burning and the respiration of animals in terms of air being "deprived of its elastic force by the breathing of animals very much in the same way as by the burning of flame." Stephen Hales, fifty years later, spoke the same language. But these men were all ahead of their times. As we reread their papers we see that in spite of strange words and ill-defined ideas they had demonstrated that air in which material had been burned or animals had respired would no longer sustain fire or life; furthermore, they showed that there was an actual diminution of the volume of the air in such cases. All of this seems to force the right explanation to our eyes; not so to the chemists of the eighteenth century. They talked in terms of phlogiston, and within limits it was a fruitful concept.

EXPERIMENTAL DIFFICULTIES WITH GASES

A chemist reading the papers of the phlogistonists clutches his head in despair; he seems to be transported to an Alice-through-the-looking-glass world. But if he is patient and interested he soon recognizes that much of the difficulty stemmed from the experimenters' inability to handle and characterize different gases. This fact illustrates the difficulty of experimentation. Metals and calxes, inflammable substances like sulfur, charcoal, and phosphorus, the chemists of the eighteenth century could recognize and manipulate since they were solids. Even some liquids like vitriolic acid, water, and mercury were quite definite individuals. But two gases, neither of which would support fire, like nitrogen and carbon dioxide, were often hopelessly confused; or two which burned, like hydrogen and carbon monoxide. Nearly all gases look alike except for the very few which are colored. They are compressible and subject to thermal expansion to about the same degree. Their densities, i.e., the weight of a unit volume, differ but that was something not easy to determine in those days. Indeed, in the eighteenth century the distinction between weight and density, even for solids and liquids, was often confused. The chemical properties of each gas are characteristic and the way each gas is prepared is different; and it was these differences that finally led to a straightening out of some of the tangled skein.

To understand the difficulties of the chemists of 175 years ago, imagine yourself an elementary student in a laboratory given glass bottles of air, of oxygen, of nitrogen, and one containing air saturated with ether vapor and

asked to tell whether or not all the "airs" or gases in the
bottles are identical. The air containing the ether vapor
(actually still largely air) will be the only one at first recognized as distinct. A student does not know how to proceed
to examine these gases except by looking at them, smelling
them, and applying a few simple tests such as noting their
solubility in water. And from Boyle's day to Priestley's the
experimenters were largely in the same predicament. They
spoke of different "airs" but hardly knew whether the differences were real or due to the presence of some impurity.
Thus Priestley, writing in 1777, said:

"Van Helmont and other chymists who succeeded him,
were acquainted with the property of some *vapours* to
suffocate, and extinguish flame, and of others to be ignited.
. . . But they had no idea that the substances (if, indeed,
they knew that they were *substances*, and not merely
properties, and *affections* of bodies which produced those
effects) were capable of being separately exhibited in the
form of a *permanently elastic vapour* . . . any more than
the thing that constitutes *smell*. In fact they knew nothing
at all of any air besides *common air*, and therefore they
applied the term to no other substances whatever. . . ."
(Priestley used the word air in the sense in which we use
the word gas today.)

The history of the study of gases covers a hundred years
from Boyle's day. A number of important improvements
in techniques were made. They were brought to a focus by
Priestley who in 1772 carried out extensive and very original experiments with "airs." He improved still further
several techniques of handling these airs or gases which
enormously simplified the experimental procedures. Prominent among these was the pneumatic trough (Fig. 25).
Before Priestley's work only three "different airs" were

known. In a few years he had discovered eleven more, including oxygen. Here is another illustration of the importance of techniques, though here we meet with an evolutionary rather than a revolutionary change.

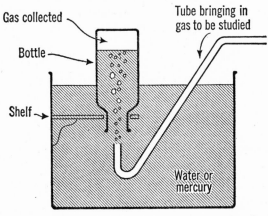

Fɪɢ. 25. Cross-sectional view of a pneumatic trough. The bottle is filled with liquid by immersing it in the water or mercury of the trough and then is raised into position on the shelf. The gas to be collected for study bubbles up into the inverted bottle, replacing the liquid.

LAVOISIER'S CLUE

The new conceptual scheme of Lavoisier, a young French amateur in science, appears to have started with his experimenting with the burning of phosphorus and of sulfur (Fig. 26). In a famous note of 1772 he wrote as follows:

"About eight days ago I discovered that sulfur in burning, far from losing weight, on the contrary gains it; . . . it is the same with phosphorus; this increase of weight arises from a prodigious quantity of air that is fixed during the combustion and combines with the vapours.

"This discovery, which I have established by experiments that I regard as decisive, has led me to think that what is observed in the combustion of sulfur and phosphorus may well take place in the case of all substances

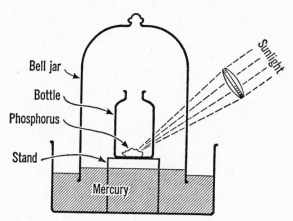

FIG. 26. Diagram of apparatus for Lavoisier's experiment on the gain in weight of phosphorus on calcination. A weighed piece of phosphorus is placed in a bottle of known weight, and the bottle is covered with a bell jar. The mercury seals off the air within the bell jar from the rest of the atmosphere. When the phosphorus is ignited (by sunlight focused with a burning glass) a white calx is formed within the bottle and the mercury level rises in the bell jar. The bottle containing the calx is removed and weighed. The calx is thus found to weigh more than the phosphorus.

that gain in weight by combustion and calcination; and I am persuaded that the increase in weight of metallic calces is due to the same cause. . . ."

Here we get a glimpse of the flash of genius involved in the origin of a new concept and how this may have been stimulated by an experimental observation. In a sense, in this note Lavoisier outlined the whole new chemistry, as he always claimed later. (The note was deposited, sealed, with the secretary of the French academy on November 1, 1772.) To be sure, Lavoisier at first mistook the gas (carbon dioxide, the "fixed air" of that day) evolved in the reduction of a calx with charcoal for the gas (oxygen) absorbed in calcination. The study we can now make of his notebooks as well as his later publications makes it plain that it was not until after Priestley's discovery of oxygen and Lavoisier's repetition of some of Priestley's experiments with this new gas that the nature of the gas absorbed in calcination became clear. It was only then that all the pieces of the puzzle fitted together, with the newly discovered oxygen occupying the central position in the picture. But at the outset Lavoisier recognized that something was absorbed from the air during calcination. Unconsciously he was retracing the steps Jean Rey had taken nearly 150 years earlier and which had never been followed up. Rey's almost forgotten book was called to Lavoisier's attention shortly after his first publication of his new theory.

An interesting question will at once come to the mind of many readers: why did the study of sulfur and phosphorus lead Lavoisier to the right type of explanation? Why after experiments with those substances did he set out full of confidence on a set of planned experiments along a new line? This is one of those historical riddles which can never be answered but concerning which it is not entirely profitless to speculate. I suggest that the key word in Lavoisier's note of November 1, 1772, is "prodigious"—"this increase of weight arises from a prodigious quantity of air that is fixed." If this is so, we have again another illustration of how experimental difficulties (or the lack of them) condition the evolution of new concepts. To determine whether air is absorbed or not during calcination of a metal is not easy;

the process takes a long time, a high temperature, and both the increase in weight and the amount of oxygen absorbed are small. But with phosphorus and sulfur the experiment was relatively easy to perform (the materials burn at once on ignition with a burning glass); furthermore, the effect observed is very large. The reason for this is that, in terms of modern chemistry, sulfur and phosphorus have low atomic weights of 32 and 31 (oxygen is 16), and in the combustion 2 atoms of phosphorus combine with 5 of oxygen, 1 atom of sulfur with 3 of oxygen. The atomic weight of the metals is high, the number of atoms of oxygen combining with them fewer. Thus 62 weights of phosphorus will yield $62 + (5 \times 16) = 142$ parts of combustion product; while in the case of tin the atomic weight is 118 and only 2 atoms of oxygen are involved. Thus 118 weights of tin would yield only $118 + (2 \times 16) = 150$ weights of calx or an increase of only about 25 per cent. Note that with phosphorus the increase is more than double. The corresponding differences would be reflected in the volume of oxygen absorbed, and furthermore, since the calcination of tin was a long process at a high temperature in a furnace, no entirely satisfactory way of measuring the volume of air absorbed was at hand in 1770.

QUANTITATIVE MEASUREMENTS AND ACCIDENTAL ERRORS

Actually, until Lavoisier was put on the track of the gas made by Priestley from mercuric oxide, he had a hard time proving that metallic calxes did gain in weight because of absorption of something from the air. The method he used was to repeat certain experiments of Boyle with a slight modification. Both the modification and the difficulties are of interest and well illustrate an important point. Boyle had sealed tin in a glass vessel and heated the vessel a long time on a charcoal fire (which he says is a very dangerous operation as the glass may well explode). He then removed the vessel from the fire and, after cooling, opened the glass, reweighed the vessel, and noted the increase in weight. This was one of the many well-known experiments showing that the calx weighed more than the metal. (Boyle, the reader will recall, believed the increase to be due to fire particles that passed through the glass.) Now, said Lavoisier, where Boyle went wrong was in not weighing the vessel *before* opening it. For if his explanation were right and the fire had passed through the glass and combined with the tin, the increase would have occurred before the air was admitted, while if oxygen were involved the increase in weight would occur *after* the air was admitted. The results obtained by Lavoisier on repeating this experiment were as he had expected but were far from being as striking as those obtained with phosphorus, for the reasons just explained. The increase was 10 parts in a total of 4,100 in one experiment and 3 parts in about the same amount in another! We now know that the difficulties of weighing a large glass vessel with a high degree of accuracy are in great part due to film moisture. It is therefore not surprising that the glass retort, after heating, varied in weight from day to day almost as much as the total gain in weight in one of the two experiments.

These tough facts of experimentation are of great importance. To me they indicate strongly that even if Boyle had weighed his vessel before and after admitting the air, the uncertainties of his figures would probably have been so great as to confuse him and subsequent investigators. *Important advances in science can be based on quantitative measurements only if the measured quantity is large as*

compared with possible systematic and accidental errors.
The principle of significant figures which played an important part in later scientific history is foreshadowed in a
crude way by this episode involving the combustion of
phosphorus and the calcination of tin.

THE PHLOGISTON THEORY:

A BLOCK TO A NEW CONCEPT

It is sometimes said that the experimenters before Lavoisier's day did not carry out quantitative experiments,
that is, they did not use the balance. If they had, we are
told, they would have discovered that combustion involves
an increase in weight and would have rejected the phlogiston theory. This is nonsense. Rey, as we have already
seen, had shown long before the beginning of the phlogiston period that a calx weighed more than a metal. Quantitative experiments, though of course not accurate by modern standards, were repeatedly made. Lavoisier in his note
in 1772 wrote as though everyone knew that a calx weighed
more than the metal from which it was formed. No straight-
forward statement of the phlogiston theory could accommodate this fact. Yet the phlogiston theory was so useful
that few if any mid-eighteenth-century scientists were attempting to overthrow it or disprove it. Rather, they were
interested in reconciling one inconvenient set of facts with
what seemed from their point of view an otherwise admi-
rable conceptual scheme.

The strategic principle which emerges is clear: it takes
a new conceptual scheme to cause the abandonment of an
old one. When only a few facts appear to be irreconcilable
with a well-established conceptual scheme, the first attempt is *not* to discard the scheme but to find some way out
of the difficulty and keep it. Likewise, the proponents of

new conceptual schemes are rarely shaken by a few alleged facts to the contrary. They seek at first to prove them wrong or to circumvent them. Thus, later, Lavoisier persisted with his own new ideas in spite of the fact that certain experiments seemed to be completely inexplicable in terms of those ideas. Only after his death was it found that the interpretation of the experiments was in error. Not so in the case of the calcination of metals: there could be no doubt in the mind of anyone by 1770 that the increase in weight during calcination was real. There was also no doubt that there should be a loss in weight according to the phlogiston theory. Or at best no change in weight if phlogiston were an imponderable substance like fire.

An early attempt to get out of the dilemma of calcination took refuge in a confusion between weight and density (calxes are less dense than metals, but the total weight in the calcination increased). This was soon put right by hard thinking. Another short-lived attempt involved assigning a negative weight to phlogiston. This illustrates how desperately men may strive to modify an old idea to make it accord with new experiments, but in this case the modification represented not a step forward but several steps to the rear! What was gained by accommodating the quantitative aspect of calcination was lost by following the consequences of negative weight to a logical conclusion. What manner of substance or principle could phlogiston be that when it was added to another material the total weight diminished? The idea that phlogiston had negative weight strained the credulity, and for the most part this logical extension of the phlogiston theory (logical in one sense, highly illogical in another) was never widely accepted. But before we laugh too hard at the investigators of the eighteenth century let us remember that before the nineteenth century heat was considered a material substance and the whole concept of the atomic and molecular theory of matter lay over the distant horizon.

The dilemma presented by the quantitative facts of calcination seems to have been accepted by the majority of chemical experimenters in the 1770's as just one of those things which cannot be fitted in. And this attitude is much more common in science than is often believed. Indeed, it is in a way a necessary attitude at certain stages of development of any concept. The keen-minded scientist, the real genius, is the man who keeps in the forefront of his thoughts these unsolved riddles. He then is ready to relate a new discovery or a new technique to the unsolved problems. He is the pioneer, the revolutionist. And it is this combination of strategy and tactics in the hands of a master which is well worth study if one would try to understand science through the historical approach.

THE EFFECTIVE DISCOVERY OF OXYGEN

It may help the reader with little or no knowledge of chemistry to outline at this point the steps in Lavoisier's development of his new ideas. These seem to have been, first, the radical notion that something was *absorbed* from the atmosphere in the calcination of metals and in combustion; second, a search for this something; third, a realization that one particular calx, mercury oxide, might provide the means for finding this something; fourth, the preparation of oxygen from mercury oxide and the failure to recognize clearly that it was not just purified common air; fifth, Priestley's publication of his evidence that the gas from mercury oxide was not common air but something new; sixth, Lavoisier's quick realization of his own faulty experimentation and his subsequent recognition that combustion and calcination involved the absorption of *a constituent* of the atmosphere, the new gas oxygen. At that point the coup d'état of the chemical revolution had taken place; from then on the rest of the drama unfolds almost as a matter of course (though, as we shall see, the acceptance of the new ideas and the abandonment of the phlogiston theory did not occur at once).

These six stages in the progress of Lavoisier's investigation from 1772 to 1777 are of great interest to anyone concerned with the mental processes of scientists of genius. It is rare that we are able to get so much insight into that complex of intuition and logical reasoning which goes into the origin of revolutionary scientific ideas. Lavoisier was a young man of twenty-nine when he began his studies of the burning of sulfur and phosphorus. It is not clear what led this wealthy man of business to take up this particular line of investigation, but the study of gases on the one hand and discussion of phlogiston on the other was a matter of current concern. He was an inexperienced experimenter, that is quite clear. Indeed, some of the claims in his first sealed note of 1772 are almost certainly incorrect; he never returned to his statements about the increase in weight on burning sulfur and one does not see how he could have made some of the observations he claimed. From this date on, however, he undertook to make himself master of the new techniques (largely developed by Priestley) for studying gases. He repeated many of the experiments of the English clergyman and also those of the famous Scotch professor, Joseph Black. From the latter's papers he probably learned to appreciate the full significance of determining the weights of the materials entering into a chemical reaction and the weights of the products.

It has sometimes been said that Lavoisier introduced the

systematic use of the balance into chemistry. This is not quite accurate, however. If any one individual is entitled to that honor it is Joseph Black. But Lavoisier insisted early in his career on the importance of weight relations. One biographer, impressed by the fact that Lavoisier was a successful member of the firm which collected the taxes for the king, suggests that the French chemist applied business principles to science. It is true that he was the first to use consistently the principle which has become an axiom of all subsequent chemists, namely that the sum of the weights of the factors (in a chemical reaction) must equal the sum of the weights of the products. This is certainly strikingly analogous to balancing a set of books; the phrase "the principle of the balance sheet" used by the biographer in question is quite apt. And the success of Lavoisier as an experimentalist (*after* the effective discovery of oxygen) as compared to Priestley's ever-increasing difficulties can be directly traced to Lavoisier's use of the "principle of the balance sheet."

In the critical stages in the unfolding of the new conceptual scheme, the balance enters only into the first observation—"the prodigious quantity of air" that was fixed in the burning of phosphorus. Our attention must be directed rather to the difficulties of interpreting experimental findings of a rough quantitative sort where gaseous volumes, not weights, were the observed quantities. But first one important chemical fact must be introduced to the reader. There was one and only one metal known in Lavoisier's time which, when calcined in air (i.e., converted to the oxide), yielded a product that at a still higher temperature reverted to the metal. This was the liquid metal, mercury. The point not only is interesting historically but underlines the importance in chemical investigations of certain materials. For a long period in the history of this science progress depended on discovering the right elements or compounds to study, just as in physics success depended on the improvement of instruments.

Priestley was probably in part responsible for Lavoisier's perception that in a study of mercury oxide lay the road to success, for Priestley told Lavoisier at a historic dinner in Paris that on heating this red powder (calcined mercury) he had obtained a gas which supported the combustion of a candle. But Priestley was under the misapprehension at that time that the gas in question was what is commonly called "laughing gas," an oxide of nitrogen which resembles oxygen in the one respect of causing a candle to burn even more brightly than in air. But with Priestley's errors we are not concerned; they are a side issue. For us, Lavoisier's errors are more important. Lavoisier, a few months after this conversation with Priestley, himself prepared a gas by heating red oxide of mercury. He examined the gas and showed that it was not carbon dioxide ("fixed air" to the chemists of the day). He took much pains on this point, for another French chemist had shortly before claimed that when mercury was formed at a high temperature from the red powder the other product was fixed air. So difficult was the early experimentation with gases.

Lavoisier, having in his hand in March, 1775, the prize he had been looking for since 1772, now made a misstep. The gas was *not* carbon dioxide; he confirmed Priestley's oral communication that "candles and burning objects were not extinguished in it, but the flame increased . . . and gave more light than in common air" (though he made no mention of the Englishman's conversation). If he had stopped right there he might at once have drawn the right conclusion, namely that he had a *new* gas before him. But he applied a test that Priestley himself had devised a few years earlier which seemed to provide a rough measure of the "goodness" of common air. That is, it gave one result with air that had been rendered "bad" by combustion or the breathing of animals and another result with common air; intermediary mixtures yielded intermediary results on a numerical scale. The chemistry of this test is too complicated for exposition in this book (it involves reactions between nitric oxide and oxygen and the absorption of the products in water). The degree of empiricism involved was certainly very high, for Priestley, talking in terms of the phlogiston theory, was essentially operating in the dark.

At all events, it so happens that when this test of Priestley's is applied *either* to pure oxygen or to common air, *the result is very nearly the same.* A diminution of volume is observed after the air to be tested has been mixed in certain proportions with the "test gas" (nitric oxide), and this decrease in volume is about equal to the volume of the test gas which had been added. There is a slight difference, but Lavoisier either failed to notice it or passed by this opportunity to become the effective discoverer of oxygen. So he reported that the gas from the calcined mercury was "diminished like common air" in the Priestley test. Therefore, his first communication to the French academy, at Easter, 1775, is confused, to put it mildly. "All these circumstances convinced me," he wrote, "that this air was not only common air but that it was more respirable, more combustible and consequently that it was more pure than even the air in which we live."

Just at the time Lavoisier was reporting to his colleagues in Paris, Priestley was busy at work also studying the gas evolved when red oxide of mercury was heated very hot.

By this time he had discovered his error in identifying the gas as laughing gas. But, exactly like Lavoisier, he was put off the right scent by his own test for the goodness of air. Then by a series of amazing accidents which it would take too long to relate he was led to examine the gas which was left after his test was complete. At once he saw that something really new was at hand. For the chemistry works out in such a way that under the empirical conditions of Priestley's test, while the change in volume is the same with common air or oxygen, what is left is entirely different: in the first case it is nitrogen, in the second oxygen. The mere introduction of a lighted candle shows the difference; and this is how Priestley first spotted the trouble.

The effective discovery of oxygen is usually considered to have taken place when in March, 1775, Priestley realized that the gas evolved from the red oxide of mercury was a *new* gas. By August of that year he had read Lavoisier's "Easter Memoir" as it was printed unofficially in a scientific journal. He realized at once the error the young Frenchman had made and pointed it out in a volume he was in the process of publishing. There seems no doubt that only after reading the announcement of Priestley's discovery and characterization of oxygen did Lavoisier realize his mistake. Then he was quick to straighten matters out. Indeed, by the time the French academy got around to printing his "Easter Memoir" the whole matter was clear. Therefore he was able to correct the first version by a few deft strokes of the pen, and as finally printed in "official" form in 1778 this "classic" paper gives no hint either of his misstep or of Priestley's unsolicited assistance. The ethics of science have changed since the eighteenth century. Today an investigator would be scrupulous about acknowledging both oral communication and prior publications.

The stumbling way in which even the greatest of strategists and tacticians have proceeded is well illustrated by this detailed story of the evolution of Lavoisier's new conceptual scheme. The rest of the narrative of the chemical revolution likewise illustrates a recurring pattern in the advance of science. A well-established conceptual scheme blocks for a time the acceptance of a new one; the conservative defenders of the old theory attempt to patch it up. Sometimes, as in the case of the phlogiston theory, the result is only a delaying action. The significant point to be noted is the way that both sides, in a controversy of this sort, put aside experimental evidence that doesn't fit into their scheme. And what is most significant, subsequent history may show that such arbitrary dismissal of "the truth" was quite justified; if so, we say there was a hidden joker in what appeared to some to be conclusive evidence.

THE LAST STAND
OF THE PHLOGISTON THEORY

By 1778 Lavoisier had made clear to the scientific world the role of oxygen in combustion. His classic experiment, often described in elementary textbooks, was as follows: Mercury heated in common air produces a red material (an oxide, we would say, a "calx" to the chemists of the eighteenth century). In a closed space about one-fifth of the air disappears in this process (Fig. 27). The red material weighs more than the metal from which it was formed. Therefore something has disappeared from the air and combined with the metal. The red material, the oxide or calx, is next strongly heated in an enclosed space with the sun's rays brought to a focus by a large lens or "burning glass," a gas is evolved, and the metal regenerated (Fig. 28). The new gas is the "something" which disappeared from the original air, for the amount is the same, and the calx has lost weight in the right amount. The new gas (oxygen) mixed with the residue from the first experiment yields a mixture which is identical with common air.

The experiments are simple, the proof appears to be complete. (Lavoisier of course generalized far beyond the case of mercury.) But the new conceptual scheme was by

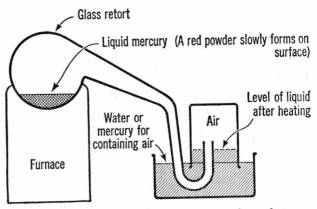

Fig. 27. Diagram of Lavoisier's apparatus to show that mercury heated in air absorbs oxygen.

no means accepted at once with great acclaim. Quite the contrary. Lavoisier had to drive home his points with telling arguments. In his *Reflections on Phlogiston,* published in 1783, he marshaled the evidence for his conceptual scheme and showed that the concept of phlogiston was unnecessary. Slowly his French contemporaries were won over, but Priestley, Watt, Cavendish, and scores of others continued to cling to the phlogiston theory. Indeed, the theory was given a new though brief lease on life by the experiments with hydrogen gas. This material was brought to the forefront of scientific discussion by Cavendish in 1766 and could be regarded as the long-sought phlogiston —or at least phlogiston combined with water. For the gas readily burned and at first no one could discover what was formed in the process. (The orthodox phlogistonists of course had the answer; the phlogiston combined with the air used in the combustion.) But just at this time the com-

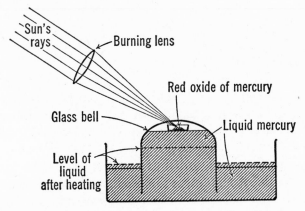

Fig. 28. Diagram of apparatus for heating red oxide of mercury and collecting the oxygen liberated.

position of water was established by the experiments of Henry Cavendish (1731–1810) which were immediately repeated by Lavoisier. Historians of chemistry still argue over who is entitled to the credit for the highly important discovery that water is composed of oxygen and hydrogen in the proportions by weight of approximately 8 to 1. Priestley, Cavendish, Lavoisier, and James Watt are all contenders for the honor.

With the discovery that water was formed when hydrogen was burned in air, Lavoisier's scheme was complete.

Water was clearly the oxide of hydrogen. Lavoisier at once proceeded to test an obvious deduction from this extension of his conceptual scheme, namely that steam heated with a metal should yield a calx and hydrogen. It did. (The converse was likewise demonstrated at about the same time.)

Hydrogen + oxygen → water
Steam heated with metal → calx + hydrogen
(oxide)

These facts about the relation of water, hydrogen, oxygen, metals, and oxides would seem to leave no ground for the supporters of the phlogiston theory to stand on. But for a few years the new knowledge had the contrary effect. The believers in phlogiston were at last able to explain why a calx weighed more than the metal. This they did by a modification of the phlogiston theory which was essentially this: instead of the calxes being treated as the simplest materials, they were regarded as compounds of water and a "pure earth." A metal was the product of the combination of phlogiston and the corresponding "pure earth."

Calcination According to Old Phlogiston Theory

Metal heated in air → calx + phlogiston (to the air)

Calcination According to Modified Phlogiston Theory

(a) metal heated in air → pure earth + phlogiston (to the air)
(b) pure earth + water from air → calx

A little study will perhaps convince the reader that these ideas accommodated the fact that a calx weighed more than its metal, for while the transformation written as (a) was regarded as one in which weight would be lost (because phlogiston left the metal), in the subsequent step (b) the absorption of water by the "pure earth" corresponded to a gain in weight. The gain in weight in this last step was arbitrarily postulated as being greater than the loss of weight in the first; the over-all result would be a gain in weight on calcination. So easily can one reconcile an old theory with new facts if sufficient ad hoc hypotheses are added with each new discovery! Incidentally this last stand of the phlogiston theory shows how vastly oversimplified is any statement that "as soon as the balance was introduced into chemistry, the phlogiston theory became untenable."

I have referred to the willingness of scientists in a controversy to by-pass alleged "facts" which stand in their way. Subsequent events sometimes show this was blind folly, sometimes inspired wisdom. Those who stood by the phlogiston theory after about 1780 had to by-pass many facts or build one ad hoc hypothesis on another. This Priestley continued to do till his death in 1804. On the other hand, Lavoisier and his followers likewise had to shrug their shoulders when confronted with certain "facts" by Priestley. They were justified in so doing as it turned out: we may perhaps say it was Lavoisier's sure intuition that led him to see which chemical transformation could be trusted to yield reliable information and which could not. Such insight or intuition has played a large and generally unrecorded role in the history of chemistry and biochemistry. Examples could be cited from our own times. But rather than attempt such an expedition into the complications of modern chemistry a few further words about Priestley's alleged facts may be in order.

In the first place, Priestley believed in many experiments,

a multitude of observations. That he was led further and further into a jungle may have its moral for workers today in fields still largely empirical. However that may be, it is of interest that while Lavoisier stuck to the one metallic oxide that yields clean-cut results Priestley was continually experimenting with other oxides and usually without quantitative control. Other oxides are hard to obtain pure, and homogeneous material is essential for a chemist. Here is our old friend the uncontrolled variable again. From a wealth of fool's gold, one might almost say, Priestley could always find some "fact" to challenge Lavoisier and support the notion of phlogiston. For example, he continued to claim that some metallic calxes yielded fixed air (carbon dioxide) on heating. So they did; but this is only because they were contaminated with the carbonates. Purity of materials is as essential to the chemist as the control of such variables as temperature and pressure is to the physicist. Adequate criteria for purity were only slowly established as Lavoisier's principle of the balance sheet became accepted. Qualitative experimentation with impure substances can lead in most cases only to confusion, and of that Priestley, brilliant tactician as he was, was often the unconscious author.

One of Priestley's main points against Lavoisier's views was based on a mistaken identification of two different gases. And that this error was not detected by Lavoisier or his followers underlines once again the difficulties of experimentation. Two gases, both inflammable, carbon monoxide and hydrogen, were at that period confused, even by the great experimenter with gases. Assuming their identity Priestley could ask Lavoisier to account for phenomena which were indeed inexplicable according to the new chemistry but could be accommodated in the phlogiston theory—now being twisted more each day to conform to new discoveries. Not until long after Lavoisier's execution in 1794 was the relationship between the two gases straightened out. Lavoisier therefore was never able to respond to the most weighty of Priestley's arguments against his doctrine. He merely ignored the alleged facts, much as Priestley ignored many of Lavoisier's experiments. Each undoubtedly believed that some way would be found around the difficulties in question. Lavoisier's hopes, not Priestley's, proved well founded. So proceeds the course of science. To suppose, with some who write about "the scientific method," that a scientific theory stands or falls on the issue of one experiment is to misunderstand science entirely.

A study of the overthrow of the phlogiston theory is thus seen to be more than a single case history; it is a related series of case histories. These illustrate the three principles to which I referred earlier in this chapter. The complicated steps in the development of a new conceptual scheme from experiments and observations are laid before us in the record: brilliant flashes, logical arguments, false steps are seen strangely intermingled. Furthermore, the study of the phlogiston theory makes one realize how old concepts may present barriers to the development of new ones. And, having traced the course of the history of experiments with gases and calcination, few would question that scientific discoveries must fit the times if they are to be fruitful. In addition, other principles of the tactics and strategy of science have constantly recurred throughout the somewhat lengthy story: the influence of new techniques, the difficulties of experimentation, the value of the controlled experi ment, the evolution of new concepts from experiment—all these are to be found illustrated in this often neglected chapter in the history of science.

THE DEVELOPMENT OF THE CHEMIST'S ATOMIC THEORY

I propose to conclude this chapter by sketching briefly the development of the atomic theory from 1800 to 1860. The significance for the general reader of this portion of the history of chemistry is twofold: on the one hand it shows how, after Lavoisier had cleared the ground and introduced quantitative methods, a conceptual scheme was required to bring order out of a mass of numerical data; on the other hand, the fifty-year lag in the acceptance of the basic ideas of the atomic theory illustrates the role of preconceived ideas or prejudice in the development of science. Indeed, if I were going to expand this section into a chapter I would entitle it "a half-century conflict of prejudices."

The historical fact that it required fifty years of experimentation and discussion to develop an atomic theory that would accommodate the experimental observations of the chemists is perhaps not surprising. But to one unfamiliar with this fact of scientific history it is amazing to discover that all the relevant ideas and all the basic data were at hand almost from the outset. An analysis of the rival views and the arguments pro and con shows clearly that certain preconceived ideas then current among scientists blocked the development. I hope to indicate in the next few pages the nature of these ideas and how the prejudices arising from them were finally overcome.

Before the reader becomes involved in the somewhat intricate mass of facts and hypotheses which will shortly follow, I venture to point up the moral of this story by a quotation from a modern writer who believes there is *a* scientific method. "The scientific way of thinking requires the habit of facing reality quite unprejudiced by any earlier conceptions." This is a fair sample of one set of views about science that is popular in many quarters, one of those half-truths with which it is most difficult to deal. If the writer means merely that science requires intellectual honesty we must all agree; if he had in mind that the scientist honestly seeks a clear-cut answer from his experiments planned within the framework of the conceptual fabric of his day and in terms of the hypotheses to be tested, then again there can be no dissent. But the statement seems to say much more than this. It implies that the mind of a scientist must be a blank when a new problem arises. Actually, as the case histories already considered show so clearly, the effective investigator must be armed with a whole series of "earlier conceptions." These are the concepts and conceptual schemes of his science, and the real pioneers must have a new idea as well. But it might be put forward in rebuttal that all these ideas are explicit; prejudices are nonlogical emotional reactions. To this I would agree and still maintain that every scientist must carry with him the scientific prejudices of his day—the many vague half-formulated assumptions which to him seem "scientific common sense." And no better illustration of the role of such factors in science can be found than in the story of the attempts of nineteenth-century chemists to formulate an atomic theory.

The discovery of the role of oxygen in combustion and the composition of water set the stage for the new chemistry. Lavoisier's *Elements of Chemistry* expounded the new ideas and impressed upon the scientific world the significance of the "principle of the balance sheet." According to the new conceptual scheme, there were two very important classes of materials: elements and compounds, the latter resulting from the combination of two or more elements in definite amounts. One could thus define water as a compound of the elements hydrogen and oxygen in the ratio by weight of 1 to 8; such a statement was the summation of difficult quantitative experimentation.

Dalton about 1805 sought to bring into the new chemistry the ancient doctrine of atoms. The idea of matter being composed of atoms was one of those general speculative ideas in the background of the thinking of most scientists of the eighteenth century; Newton had made use of the concept of atoms in some of his writings about the physical properties of gases. To Dalton goes the credit, however, of suggesting that the atoms would provide a simple explanation of why the elements always combine in the same proportion by weight to form a definite compound. If one assumes that all the atoms of a given element, say hydrogen, have the same weight and that the combination of, say, hydrogen and oxygen always involves the union of the same number of atoms, then the experimental facts are adequately accommodated. To illustrate, let us make the simplifying assumption (as did Dalton) that the smallest water particle is composed of one atom of hydrogen and one of oxygen. (According to our present conceptual scheme this is incorrect.) Then since we know by experiment that 1 part by weight of hydrogen combines with 8 parts by weight of oxygen, the *relative* weights of the hydrogen and oxygen atoms are 1 to 8. Though atoms were far too small to weigh individually, Dalton argued one could determine their relative weights from the experimental facts by the type of reasoning we have just employed.

There was one difficulty apparent at the start and this difficulty remained to plague the chemists for half a century. How is one to know how many atoms unite to form the smallest unit of a compound? Dalton said we cannot know but had best *assume* the simplest relationship which will fit the experimental facts. One may note that here is an example of invoking the general principle which has sometimes been called the "rule of greatest simplicity." Dalton said the water molecule (our name for the smallest particle of a compound) was composed of an atom of oxygen and one of hydrogen; in modern symbols his formula for water was HO. This assumption made, *then* it follows from the weights of hydrogen and oxygen which combine that an atomic-weight scale can be constructed in which, if hydrogen is taken as unity (arbitrarily), oxygen must be 8.

Briefly, the scientists in the first half of the nineteenth century were wrestling with a threefold relationship of which only *one* point was fixed by experiment, namely the combining weight ratios of the elements. If one assumed a formula for a few compounds such as water, then an atomic-weight scale could be constructed from the experimental facts; or, conversely, if one assumed an atomic-weight scale, then the formulas of the compounds followed from the weight relationships of the elements. But what was needed was new evidence as to either the relative weights of the elementary atoms or the number of atoms in such simple compounds as water.

Both the evidence and new concepts for interpreting the evidence were presented to the scientific world in the second decade of the nineteenth century. *But this fruitful combination was ignored.* Avogadro, an Italian physicist, saw how another set of quantitative data could be used to establish the formula of water. But it was not until 1860 that, under the leadership of another Italian, Cannizzaro, the scientific fraternity returned to the ideas of Avogadro and built on them an atomic and molecular theory that could be widely accepted. Yet this is the theory which has served so well as a foundation for all subsequent developments about the structure of matter.

To understand the prejudices which prevented Avogadro's ideas being accepted in his own day we must examine both the experimental evidence and his concepts. Avogadro (and some of his contemporaries) were impressed by the quantitative relationship that had been found to hold when *gaseous* elements combine. Here we are considering not weight but volumes of gases, let it be carefully noted. One illustration will suffice. If a mixture of hydrogen gas and oxygen gas is exploded by a spark, the following relationship is found to hold:

1 volume of oxygen + 2 volumes of hydrogen → 2 volumes of water vapor.

(Any units of volume, for example cubic feet, may be used to express this relation.) The relation between volumes is very simple: 1 to 2 to 2. Other gaseous elements were likewise found to combine in a volume relationship expressed by *small whole numbers.*

Avogadro made two assumptions to account for the whole-number relationship between the *volumes* of gaseous elements which combine to form compounds. The first was that *equal volumes of gases under the same conditions of temperature and pressure contain the same number of particles.* The second was that the particles of hydrogen and oxygen are each composed of *two* atoms united together.

With the aid of these assumptions Avogadro accounted

for all the known facts about chemical reactions between gases and was led to the conclusion that the water molecule was composed of two atoms of hydrogen and one of oxygen, that is, it is to be represented by H_2O. Thus, by bringing in as new evidence the measurement of the *volume* relationship between *gaseous* elements which combine and his own chain of reasoning, Avogadro established the formula of a few compounds. He thus solved the threefold relationship referred to earlier; the atomic-weight scale could then be constructed from the proportions by weight in which the elements combined. To be specific, the formula of water as

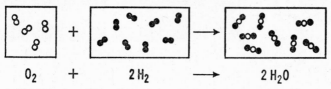

$$O_2 \quad + \quad 2\,H_2 \quad \longrightarrow \quad 2\,H_2O$$

Fig. 29. Diagram illustrating Avogadro's view of gaseous combination; the square represents one volume, each oblong two.

H_2O followed from the *volume* relationship between water vapor, hydrogen, and oxygen in the synthesis of water from the elements; the atomic weight of oxygen must then be 16 if hydrogen is 1, since the weight relations of oxygen and hydrogen in water are as 8 is to 1 or 16 is to 2. The combination of oxygen and hydrogen Avogadro pictured as shown in Fig. 29 (and this is the way we picture it today).

Dalton would have none of this; neither would the vast majority of chemists for nearly fifty years. Why? Because Avogadro's conceptual scheme involved the assumption that either the particles of hydrogen gas were divisible, which was contrary to the definition of an atom, or else the particles were composed of two *like* atoms held together. But what could hold two identical atoms together?

Berzelius, the great Swedish chemist of the first half of the nineteenth century, was particularly insistent that identical atoms would not unite, for he had developed a concept of chemical combination that depended on the assumption of an electrical attraction between different kinds of atoms. This was a conceptual scheme which reflected the recent discovery of the decomposition of some compounds such as water by an electric current (electrolysis). According to Berzelius' electrochemical ideas, identical atoms could not combine.

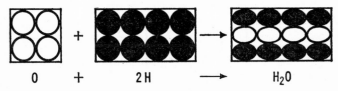

$$O \quad + \quad 2H \quad \longrightarrow \quad H_2O$$

Fig. 30. Diagram illustrating Berzelius' view of gaseous combination.

Berzelius, however, accepted half of the Italian physicist's postulates. He said that the fact that one volume of oxygen combines with two volumes of hydrogen is of the utmost significance and that for *gaseous elements* Avogadro's premise is undoubtedly correct (i.e., equal volumes of gaseous elements contained an equal number of atoms). Therefore Berzelius formulated the combination of oxygen and hydrogen as shown in Fig. 30.

In this diagram I have represented the atoms as touching each other, for Berzelius and his contemporaries thought of the particles in a gas as contiguous. In Avogadro's conceptual scheme we think of the molecules as occupying but a small portion of the space. Berzelius did not account for the fact that exactly two volumes of steam were formed. In his picture the atoms in the water molecule were squeezed together, but why the volume of the product was related so simply to that of the components he did not say. Perhaps it was "just one of those things we cannot yet explain." In spite of its inadequacies this conceptual scheme was very fruitful. With his assumptions Berzelius developed a system of chemistry of great value, but before long it ran into difficulties and was discarded.

In retrospect, we see that there were at least three prejudices current in 1815 unfavorable to Avogadro's ideas. The first was the notion that the particles of a gas were in contact with each other; the second, that identical atoms could not unite; a third was Berzelius' electrochemical theory. The latter was so explicit an idea, however, that it is perhaps wrong to call it a prejudice; rather it was a conceptual scheme which served to block another's acceptance.

One attempt to obtain experimental evidence to support Avogadro's ideas was made by a French chemist in the 1820's, but the result of these studies was almost to destroy all faith in the atomic theory. This is indeed a curious chapter in the history of nineteenth-century science. In a few words, what occurred was this: a way was discovered of measuring the relative weights of the vapors of such elements as mercury and sulfur, which are gases only at high temperatures. If Avogadro's first postulate is sound, the relative weights of equal volumes of gases should be a measure of the relative weights of the particles which compose these gases (since the assumption was that equal volumes contained an equal number of particles). The followers of Berzelius accepted this idea provided one considered only the elements. Now came the disconcerting discoveries. The relative weights of vapors of most of the elements could not be reconciled with the data used for constructing the atomic-weight scale. In modern terms one would say that if hydrogen gas were to be represented as H_2 (two atoms per molecule), in gaseous mercury the atoms were single, while gaseous sulfur had to have no less than six atoms united together.

Such conclusions seemed preposterous to the chemists of the 1830's. The simple gaseous elements, hydrogen, oxygen, nitrogen, chlorine, all behaved in a uniform manner (they could be regarded as having either one atom per particle according to Berzelius or two per particle according to Avogadro). Who could imagine that nature was so contrary as to have other elements with one or six particles per molecule? Often, unconsciously, scientists operate on the principle of assuming the greatest simplicity in nature. The application of this principle to the elements in the first few decades of the nineteenth century certainly led to the idea that the number of atoms per particle in the gaseous state should be the same for the elements. Therefore, confronted with the alternative of throwing over the principle of uniformity among the elements or abandoning Avogadro's postulate, most scientists took the latter course. With the postulate likewise went most of Berzelius' system (which was in trouble on other counts as well). Therefore the 1840's saw a period of reaction against the whole atomic theory. Any attempt to find a basis for determining the number of elementary atoms united in a compound molecule seemed hopeless. Chemists retreated to a position similar to that taken by Dalton. They arbitrarily assumed the law of the greatest simplicity and wrote the formula of water HO and a corresponding atomic-weight scale.

It would be interesting, if space permitted and if the necessary factual information were not so great, to trace the ebb and flow of the faith of the scientific world in the "reality" of atoms. The 1840's and 1850's would represent a very low point indeed. But as a consequence of the development of the kinetic theory of gases to account for physical phenomena, Avogadro's first postulate came to seem more and more plausible. Furthermore, without an adequate atomic theory further progress in chemistry was blocked. The complicated facts of the chemistry of organic compounds (carbon compounds) could not be handled without some agreement as to the number of atoms in a molecule of at least the simple substances. A generation of chemists and physicists produced a mass of converging evidence. Then in almost a rush everyone turned back to Avogadro. The earlier doubts were overcome: a great many facts revealed by the workers of the forties and the fifties were seen to fit into Avogadro's conceptual scheme. Unlike the situation when Lavoisier was fighting for his ideas, there was no rival doctrine, only a state of confusion. Therefore in 1860, largely as the result of a beautiful exposition by Cannizzaro of Avogadro's ideas, the atomic-molecular theory was accepted in the form in which it is taught to elementary students of chemistry today. Immediately there followed rapid progress in chemistry in many directions. This illustrates again the revolutionary effects of concepts or conceptual schemes which like new instruments often open up wide fields for exploration. But that is another story. Now that the reader has come to the mid-nineteenth century in the history of chemistry it is time to turn to other sciences to illustrate other patterns of scientific work.

CHAPTER EIGHT

The Study of Living Organisms: Natural History and Experimental Biology

A STUDENT OF biology might well complain that up to this point I have been discussing not experimental science but merely the physical sciences. The point would be well taken. This chapter and the next represent an attempt to right the balance. However, the experimental biologist of today is so deeply concerned with chemical and physical problems that a thorough understanding of the physical sciences is a sine qua non for an understanding of biology. This is certainly true of the investigator, and I believe it is no less true of the onlooker. Visit almost any laboratory in a medical school, a hospital, or an agricultural station; examine the equipment and chat with the men and women at work. You will have difficulty in distinguishing what is going on from what might be observed in a laboratory of chemistry or nuclear physics.

There would be, indeed, one distinguishing characteristic: almost without exception an explanation of the problem under investigation would reveal some reference to a living organism. Indeed, there might well be plants or animals right in the laboratory or, if not, certainly near by. A clinic, a greenhouse, an animal house, or an experimental farm would directly or indirectly fall within the frame of reference of the investigator. Whether the research worker call himself a clinician, a medical scientist, a plant physiologist, a biochemist, or a biophysicist, if the orientation of his thoughts is toward a living entity we may place his labors in the broad category of experimental biology. To a large degree the patterns of investigation and the principles of tactics and strategy of science developed in the preceding chapters are relevant to an understanding of this vast and highly important field. There are, however, certain special considerations in this whole area of investigation of which the reader should be made aware.

In the first place, the historical development of systematic biology must be related to the work of the experimental biologist. In the second place, current studies in what may be called observational biology must be examined as to methods. For the elucidation of the life history of organisms, the geographical distribution of plants and animals, and the further elaboration of schemes of classification are today major sectors of biological inquiry. The methods of the systematist and student of natural history seem at first sight so closely related to common-sense procedures as to represent a different type of activity from those we have been considering. Indeed a few scientists working in these fields might be inclined to quarrel with my definition of science; they may prefer to define science as systematized knowledge. Yet Marston Bates in a recent book, *The Nature of Natural History*, points out that "classification is essentially a conceptual scheme," and goes on to say that any attempt to make a classification rigid and inflexible removes it from the area of science. For this biologist, at least, the definition of science as "an interconnected series of concepts and conceptual schemes arising from experiment and observation and yielding new experiments and observations" is broad enough to include all of biology. At the outset of a consideration of some characteristics of experimentation in biology, however, it is worth while stopping a moment to amend slightly some of the earlier statements in this book about the development of modern science.

In identifying the three streams of thought and action (p. 45) which came together in the sixteenth and seventeenth centuries to make modern science, only passing reference was made to agriculture and to medicine. The age-old tradition of cut-and-try experimentation was presented primarily in terms of metalworkers and similar artisans. Yet clearly the procedures which have been developed over the centuries for growing plants, breeding domesticated animals, making fermented beverages, baking bread, and preparing all manner of foods are excellent examples of pure empiricism. In an address delivered in 1876 John Tyndall used the word empirical in exactly the sense in which I have been employing it in the preceding pages, and he was speaking of biological phenomena. "Until the present year," he said, "no thorough and scientific account was ever given of the agencies which come into play in the manufacture of beer, of the conditions necessary to its health, and of the maladies and vicissitudes to which it is subject. Hitherto the art and practice of the brewer have resembled those of the physician, both being founded on empirical observation. By this is meant the observation of facts, apart from the principles which explain them, and which give the mind an intelligent mastery over them. The brewer learnt from long experience the conditions, not the reasons, of success."

To the long experience of those concerned with the preparation of food and drink must be added the *empirical observations* of the early anatomists and naturalists. These observations resulted, however, not so much in modification of practical procedures as in the development of ideas about the structure and relationships of living organisms. We must, therefore, consider amending not only our characterization of the experimental tradition but also that of the two streams of thought—speculative ideas and deductive reasoning.

In antiquity and the Middle Ages the classification of animals and plants and a description of the structure of many types were an important part of the learned tradition. Aristotle's writings on natural history served for long as a model of how the manipulation of general ideas by logical processes can yield classificatory schemes. Aristotle was both a careful observer of nature and a logician of towering strength. Even after some of his statements about natural history began to be questioned, the influence of his methods continued to be powerful. To bring the biological sciences into proper focus, therefore, the stream of thought I have designated "deductive reasoning" must be broadened to include the logic of categories and classes.

With these relatively minor amendments my earlier general description of the origin of modern science in the sixteenth and seventeenth centuries would seem to apply as well to the biological as to the physical sciences. In biology the new concepts were related for the most part to observations instead of to experimentation; the general ideas were developed to improve the description of structure and relationships.

In one important respect, however, the history of the biological sciences differs from the history of the physical sciences. The biologist is never able to withdraw himself as far into an artificial world of experimentation; his ability to experiment with artifacts is far more limited than that of his neighbors in the physical sciences. Common-sense considerations and practical problems are always close at hand. His sphere is limited by definition to living organisms (though he may examine dead ones to elucidate certain problems); that means he must work with the material presented to him by nature. Perhaps a penetrating philosophic analysis would show that the distinction I am making is more apparent than real. For one may say that while the chemist can create a world of synthetic compounds that never existed before, he is only reshuffling combinations of materials presented to him by nature; even the physicists who in our time have made new elements can be regarded as merely realizing the potentialities inherent in matter. Yet looking at the matter historically I would still maintain that there is an important difference. The biologist, however much he may wish to pursue "science for its own sake," is concerned with living entities and these same entities are of prime importance for the material welfare of the human race. As a consequence, in the development of biology as a science there never could be as much of a divorce between theory and practice as was for long the case in physics and chemistry. Except for the navigator, the practical man of the first half of the eighteenth century took relatively little advantage of the recent advances in physical science. Much as the founders of the Royal Society sought to make their science useful, their success was highly limited.

The contrast I am drawing between the physical sciences and the biological sciences in the period 1550–1850 is the contrast between an increasingly theoretical and abstract group of sciences largely divorced from practice on the one hand and a consistently empirical endeavor tied closely to practical considerations on the other. Systematic biology is nonmathematical and the concepts employed are very like those of everyday life. Until the nineteenth century broad working hypotheses and specialized concepts played relatively little part in the growth of the biological sciences, whose progress can be measured largely in terms of accumulated information. The systematization of a mass of empirical observations was highly prized, however, because of a fundamental aspect of biology, namely, its overriding importance to human beings. Sickness and death have never been far removed from people's thoughts; medicine, perhaps, ranks as the oldest of the learned professions. Therefore, when in the Renaissance there was a renewed interest in firsthand observation, the schools of medicine became centers of the new science. The professors at Padua in the sixteenth century explored the anatomy of the human body; they did so with the hope of relating their findings to the practice of medicine. It is interesting to note that William Harvey was studying anatomy with Fabricius at Padua in 1600 when Galileo was a professor at the same university.

Harvey's discovery of the circulation of the blood in 1628 is one of the landmarks in the history of science. The new concept of the blood as a *circulatory* fluid arose from his observations and experiments and has certainly been fruitful beyond measure. One may note in passing my use of the word concept; the reader may be inclined to say the circulation of the blood is not a concept but a "fact." When the idea was put forward it was as much of a new concept, however, as the idea of a "sea of air," and I shall not repeat here what was said in an earlier chapter about the difficult questions involved in all attempts to differentiate between hypotheses, concepts, and "matters of fact." Of course, the concept of the function of an animal organ such as the heart is far less general and abstract than the concept of a sea of air; and both are much nearer common-sense ideas than such a conceptual scheme as the atomic theory or the concept of a caloric fluid. But Harvey was combining the tradition of empirical observation and experiment with an interest in general principles quite as much as was his contemporary, Pascal. In his hands the observational accuracy of the Paduan anatomists led to a highly significant generalization, and although his conclusions were qualitative and could not be cast in even approximate mathematical form, no one will deny their relationship to contemporaneous progress in physical science.

Experiments designed to throw light on the question of spontaneous generation provide another example of the penetration of experimental methods into biology as early as the mid-seventeenth century. Francesco Redi, a member of that famous Accademia del Cimento of Florence referred to in earlier chapters, was a physician who was interested in the experimental science of his day. He combined careful observation with cut-and-try experimentation in an endeavor to obtain an answer to a general and hence scientific question, namely, do certain types of organisms found in putrefying flesh arise spontaneously? A consideration of Redi's experiments belongs, however, in the next chapter.

Let us pursue still further our brief survey of what I might venture to call "observational biology." Various schemes for classifying plants and animals must have arisen early in the history of mankind. Likewise, rules for identification of plants and seeds must have become a necessity for those who lived by agriculture. Thus, by the dawn of written history man was in possession of a vast fund of knowledge based on empirical observations of biological phenomena; this cultural treasure was literally vital to his existence. One is hardly justified, however, in calling this store of information scientific knowledge. I should designate it an array of purely empirical data. But when learned men with philosophic minds became interested in animate nature, this mass of practical information became a subject for analysis and speculation. Some order was brought into the scheme and interest was aroused in observation of nature just for the sake of adding to the systematized knowledge. In the Middle Ages Aristotle was the authority on the morphology and habits of animals; Galen held the corresponding place as regards human anatomy and physiology. Were their writings scientific? If so, when did systematic biology become a science? If not, can we dismiss as non-scientific the writings of Fabricius, the last of the line of Paduan anatomists and the teacher of Harvey?

One can debate such questions for many days and with little profit. I suggest that a more fruitful method of examining the history of biology is to bring in a notion which has proved useful in analyzing the relation of science to industry. It will be recalled that it is convenient to speak of the "degree of empiricism" present in a given practical art at any date, and also to use the same phrase in regard to a field of pure science (pp. 56, 58). The same basic idea may be applied to a mass of data. To the extent that some general ideas and specialized concepts are introduced to provide the framework for classification, the degree of empiricism is reduced. But unless there is reference outside of the classification to still broader and more abstract ideas, one must regard the systematized knowledge as highly empirical.

To illustrate the point at issue, let me digress a moment by considering not animate but inanimate nature. What is now called mineralogy and petrography has grown out of very primitive schemes for classifying rocks and earths. This information was, of course, essential for mining and metallurgy. If one examines the mining handbooks of even the eighteenth century, it will be evident that the information is essentially empirical and yet extremely valuable for the makers of metals. Only after the chemical revolution of the late eighteenth century was it possible to put mineralogy on what we now call a rational basis. Indeed, the first great exploiter of the atomic theory, the chemist Berzelius, directed his attack on mineralogy with the aim of making it a true science. Because we have long classified the mineral kingdom in terms of the concepts and conceptual schemes of chemistry and physics, we are inclined to dismiss early mineralogy as completely empirical. Yet there are a few rudimentary concepts and general ideas even in the seventeenth- and eighteenth-century handbooks. Therefore one could trace a slight reduction in the degree of empiricism in mineralogy even before the enormous reduction caused by the chemical revolution and the advent of the atomic theory.

It is probably fair to say that in 1850 systematic biology was comparable in status to mineralogy in 1750. Even today one may doubt whether the equivalent of the impact of the atomic theory has been felt, although the introduction of the hypothesis of evolution in the nineteenth century and the rapid progress in genetics in the twentieth century have altered the picture profoundly. One would hardly question that those who now devote their lives to systematic biology and the study of the life histories and relationships of plants and animals are operating in a realm where the degree of empiricism is high. Yet until recent years few doubted the significance of adding still further information to the mass of empirical observations that is recorded in the handbooks of biology and zoology and represented by the specimens in herbariums and museums.

The reason for the continued interest in a field of activity which has remained so largely empirical is related, I believe, to both practical and sentimental considerations. If factual information is considered important, then any system of classification, however arbitrary, is better than no system. A deep concern with living organisms seems to me to have been the predominant driving force in the development of natural history and systematic biology. The practical reasons that stem from agriculture and medicine have already been considered. But over and above this there is another reason: people feel themselves far more involved with the animate world than the inanimate one. One seems closer to something important when pondering on the relationships of plants and animals than when attempting to classify minerals without knowledge of chemistry. Furthermore, common-sense observations and common-sense concepts seem to yield more information about living organisms than about minerals. At all events, from the Renaissance on, explorers and naturalists have combined to fill museums with specimens of plants from all over the world, and the activities of the scientific descendents of these early biologists have continued to the present day.

With the invention of the microscope in the seventeenth century a whole new world was opened for exploration. The improvement of the instrument in the early nineteenth century still further extended the range and made possible a study of the life history of microorganisms. The bearing of this work on the controversy about spontaneous generation will be apparent shortly. The improved microscope likewise made possible a study of the structure of plant and animal tissue; microscopic anatomy extended gross anatomy. Parallel with the continued interest in classification and the identification of new species or varieties of all manner of plants and animals went a more careful study of the process of reproduction and an elucidation of the complex life histories of many organisms. Since in many cases the small plants or animals were agents of disease of larger organisms, the medical man, the veterinary, and the applied botanist were moved to pursue this line of inquiry with special vigor. For an interesting account of what has been accomplished, the reader may be referred to the excellent book of Marston Bates previously mentioned, *The Nature of Natural History.*

Whether or not the work of the systematist in biology has reached the point of diminishing returns is debatable. To some degree there is a marked cleavage today within the biological fraternity between the taxonomist (the classifier) and the experimental biologist; the student of the life habits and relationships of plants and animals occupies an intermediate position. The rapid progress in the fields of genetics, cytology, physiology, biochemistry, and other experimental fields seems to indicate that revolutionary advances in biology are in the offing. If so, these may be reflected before long in the systematic field in much the same way that the advances in chemistry of the early nineteenth century profoundly affected mineralogy. But for the onlooker it is perhaps sufficient to be aware of the role that observational biology has played in the past and to know something of its relationships to experimental biology, which is certainly the rapidly expanding field at the present moment.

Before leaving the subject of systematic biology, a few words in general about systematized knowledge may be in order. Whatever the future may hold in store for the next generation of curators of zoological museums and herbariums, I feel that this group of scientists has unconsciously done science a poor turn by placing classification and factual information in the foreground of the picture. For the reasons I have given, a listing of all known species, sub-species, and varieties of plants and animals may be regarded as an important undertaking. (Though I have a suspicion that filling in some of the remaining gaps in certain instances is hardly significant scientific work.) But to rationalize this activity on some other than practical grounds has resulted in the doctrine that "science is systematized knowledge." This in turn leads to the conclusion that the discovery of any new bit of knowledge that can be fitted into any system constitutes an advance in science. One has only to turn to the field of organic chemistry to show the absurdity of this position.

The number of compounds of carbon that can be prepared by an organic chemist in his laboratory appears to be almost limitless; certainly the upper limit would be a figure of astronomical magnitude. Let me illustrate by one example. The simplest group of carbon compounds is that in which each member contains only carbon and hydrogen. According to the conceptual scheme of modern chemistry, we say that each member of a series of the compounds called "paraffin hydrocarbons" contains carbon atoms and hydrogen atoms in a certain relation. The series starts with a compound with one carbon atom and four hydrogen atoms; we therefore write the formula CH_4. The next member is represented by C_2H_6, the next C_3H_8, then follows C_4H_{10}, C_5H_{12}, C_6H_{14}. Now these formulas do not always represent a single substance. In fact, except in the case of the first few members they never do. There are two different compounds which have the formula C_4H_{10}, three with the formula C_5H_{12}, five represented by C_6H_{14}, and nine substances all of which must be represented by the same general formula with seven carbon atoms and sixteen hydrogen atoms. The number of these *isomers,* as they are called, has corresponded to the number predicted by the conceptual scheme used by organic chemistry. No one has any doubts as to the adequacy of the theory in this respect. Yet for the higher members of the series no one has bothered to try to prepare all possible isomers; indeed, the task would be overwhelming. For example, it can be calculated that there are over 300,000 isomers represented by the formula $C_{20}H_{42}$ and about 70,000,000,000,000 isomers with the formula $C_{40}H_{82}$. And this is only one series of relatively simple carbon compounds!

No one has attempted to calculate the number of possible carbon compounds containing from one to forty carbon atoms and hydrogen, oxygen, and nitrogen in all manner of possible combinations. But it is quite clear from the preceding figures that one cannot easily envision the day when all the theoretically possible organic compounds with even as few as ten carbon atoms will have been synthesized. Every new substance that is prepared, characterized, and whose structure is determined, represents a definite increment of new knowledge. An article describing a dozen or so new compounds will be published in the most respectable journal. But one may well ask, does such a paper represent a significant advance in science? How does such activity differ from stamp collecting? This question is worth dwelling on for a moment for it has implications that extend far beyond the province of the chemistry of carbon compounds.

Tolstoy somewhere in his writings condemns and ridicules science as a useless activity concerned with "counting the number of lady bugs on the planet." (This was before the effects of the by-products of science on industry and medicine had become apparent.) Scientists on being confronted with such statements are quick to defend their labors against such gross misrepresentation. Yet the attitude which lies behind the humanitarian's impatience with "mere intellectual curiosity" is worthy of attention. One cannot answer dogma by dogma. There is no use in the scientist's replying to a derision of "useless knowledge" by repeating the famous toast, "Here's to pure mathematics, may it never be applied." The scientist himself may well nail his flag to the mast with the equivalent of the motto "art for art's sake," but so too can the big game hunter, the mountain climber, and the stamp collector. If we are going to consider pure science as something other than a private pastime we must face up to all who question its significance.

Emphasis on science as systematized knowledge seems unfortunate. Not that such knowledge is not part of the fabric of science but it is not its essence. Such knowledge is of vast practical importance to those who work with animate and inanimate nature. But only to the degree that new *ideas* are introduced into the systematization of factual information does the activity partake of the nature of the work of those investigators we readily identify as "great scientists."

After this digression on organic chemistry, a further paragraph or two on the subject of natural history is in order. If one turns the pages of a college text on systematic zoology or biology (or takes an orthodox course in either field) one is impressed by the vast array of factual information. The emphasis is usually on the information, not as an end in itself, of course, but as related to the function and development of plants and animals and their relation to one another. Another way of looking at this impressive collection of data is to consider how in the world it ever came into being. This is the aspect of biology which can be found only in the histories of this subject. But it is here that one sees the methods of observation employed by the pioneer biologists and realizes the significance of new results as the stimulus for further work.

I refer the reader to the last chapters of Wightman's *The Growth of Scientific Ideas* and to histories of biology for an understanding of the growth of the biological sciences. Here I should like to emphasize only a few points to connect the methods used in biology with what has been said earlier about the physical sciences.

In the first place, the significance of a new instrument and of the improvement of an old instrument is written in large letters throughout the story of biology from the seventeenth century on. It is so obvious, however, that biologists often fail to place it in proper perspective in writing for the layman. Microscopic observations started in the seventeenth century; indeed, the last fifty years of that century are often spoken of as the age of the classical microscopists. (Interestingly enough, one of the greatest of the heroes of this age, the Dutch microscopist Anthony van Leeuwen-hoek, worked with simple lenses—a magnifying glass rather than what we usually consider a microscope.) The second great age of microscopy came in the second quarter of the nineteenth century. This seems to have been largely due to the fact that difficulties with the use of two lenses (the com-pound microscope) were then finally overcome. This was the consequence of, first, the formulation of an adequate conceptual scheme accounting for the behavior of light of different colors when passed through a lens and, second, the practical discovery of how to make different kinds of glass and combine them. The result was an instrument not far different from the modern microscope which gives a high degree of magnification and a satisfactory image be-cause light of different colors is brought to the same sharp focus. Furthermore, by the middle of the century new techniques of viewing objects and preparing cross sections of specimens had been well developed and were widely used.

The classical microscopists of the seventeenth century revealed a new world of living organisms. Some of the consequences of this will be evident when we consider the history of spontaneous generation. They likewise discov-ered the spermatozoa in animal semen and by this and similar observations took important steps along the road of discovering the complexities of the phenomena of repro-duction. But it was not until the nineteenth-century biolo-gists, armed with really potent microscopes, came into action that it was possible to understand many intricate processes such, for example, as those involved in the repro-duction of the flowering plants.

The second point to be made is that in exploring the structure of the development of living organisms the plan

of campaign is almost identical with such occurrences as
finding a boy lost in the woods or tracking an animal. Both
as to procedures and as to concepts, natural history and
systematic biology in spite of the vast complexity of detail
are very close to common sense. Indeed, in this whole area
my cautious and skeptical approach to scientific concepts
and conceptual schemes may seem out of place. One may
attempt to believe that it is not a "fact" that we "live on a
globe surrounded by a sea of air," that this is merely a
highly probable conceptual scheme; but it would be a
hardy skeptic indeed who denied that it was a "fact" that
the fertilized egg resulted from the union of a spermato-
zoon with an ovum, to choose one from a million examples.
Or consider the process of pollination: it seems almost un-
believable to one who has never studied botany and comes
for the first time on the description of the details. Microscopic observations, we are told, show that from the pollen
grain there grows a tube which descends the style and enters
the ovary. By this process a generative cell from the pollen
unites with one from the pollinated flower. Complex, admittedly, but certainly factual. Or to choose another example, we may marvel at the detective skill of those who
have traced the strange life history of certain parasites of
domestic animals and man; and we seem to be dealing with
a procedure akin to that of the F.B.I. when we follow the
naturalist disentangling a complicated mass of facts.

In short, the concepts in systematic biology at present
and the procedures used by the fieldworker are so similar to
everyday activities that in this area of science we seem to
be very close indeed to common sense. But there are areas
of biology today where the concepts and conceptual
schemes are for the nonscientists as remote from common
sense as those of the physicists and chemists. One of these

is the field of genetics. The concept of the gene (the unit of heredity) is in origin at least as hypothetical as was the concept of the atom when introduced into chemistry by Dalton around 1800. This is an example of a biological concept; or one might better say the modern formulation of the mechanism of heredity is a biological conceptual scheme. Some notion of how these new ideas have developed and of the present status of the conceptual scheme are of special interest at the present moment. For it so happens that within the Soviet Union the basic ideas accepted by almost all geneticists today have been declared incompatible with the doctrine of the Communist party. Ideology and science have become strangely mixed. Of this I shall have more to say in the concluding chapter of this book. But I recommend Julian Huxley's book, *Heredity, East and West*, both as an introduction to this branch of biology and as an interesting account of what has happened to science the other side of the Iron Curtain.

PASTEUR'S STUDY OF FERMENTATION AS AN ILLUSTRATION OF EXPERIMENTAL BIOLOGY

Another field of biology in which at once one meets concepts and conceptual schemes far distant from common-sense ideas is physiology. Here the reason is primarily because the study of the life processes of plants and animals has led investigators over the last century and a half deeply into physics and chemistry. Therefore the whole conceptual fabric of these sciences has penetrated biology along an ever-widening front (including the study of heredity, we may note in passing). To illustrate how this has occurred as well as to show some of the difficulties of applying the logic and methods of chemistry to biology, we turn to Louis Pasteur's study of fermentation.

First of all, anyone at all interested in the methods of science should read the biography of Pasteur by Reneé J. Dubos, *Louis Pasteur, Free Lance of Science*. In a chapter entitled "From Crystals to Life" Doctor Dubos gives a lucid description of how a young French chemist in the middle of the nineteenth century started by studying crystallography and ended by studying living organisms. Between the ages of twenty-three and thirty-three Louis Pasteur shifted his attention from inanimate to animate nature. The factors influencing this reorientation of a genius can be discovered fairly readily from his writings and those of his contemporaries; they are not without interest for historians of science and have special significance for anyone who attempts to write on the tactics and strategy of science.

It is clear that Pasteur was led in part to his study of fermentation by his interest in a practical problem, the alcoholic fermentation of beet sugar in a distillery. He was called in for consultation by a distiller in Lille where he was dean of the Faculty of Science in the university. Here we see reflected a familiar pattern already referred to more than once. A technical matter arouses the interest of a scientist and leads to a development in pure science. But quite apart from this chance contact with the art of fermentation, Pasteur had other reasons for undertaking an investigation of the biological processes called fermentation. As a consequence of his study of crystals he had developed a far-reaching hypothesis, and it is to this that he refers in the opening of his first paper on fermentation. The hypothesis was bold and with certain modifications stands as an unchallenged generalization today. At the time of its formulation, however, the evidence was very slight; but a fruitful hypothesis it certainly proved to be.

Pasteur had been studying the rotation of the plane of

polarized light by certain crystals and liquids. This optical phenomenon is somewhat complicated and for present purposes may be regarded as a physical characteristic which can be measured by the use of suitable instruments. The property of affecting polarized light in the manner described by scientists by the word "rotation" is relatively rare among liquids or substances in solution. (I am purposely omitting a long and fascinating story about Pasteur's work on crystals.) The materials which have this property are all products of the plant and animal world. Today we have many instances to support the generalization that only products of life produce the effects in question. The number of examples known to Pasteur were few but nevertheless he became firmly convinced *that only materials formed as a consequence of a life process would rotate the plane of polarized light.* Here was his far-reaching hypothesis. Therefore, when he was confronted with the fact that a certain material, amyl alcohol, a by-product of lactic acid fermentation, had the ability to rotate the plane of polarized light, he concluded that some living organism must have been involved in its production.

I have simplified the story, for a reading of Pasteur's paper shows that in addition to this grand hypothesis another was likewise in his mind. But the other hypothesis cannot be explained without a long digression into technical matters, and as it turned out, this other hypothesis was soon found to be quite invalid. The important point is that Pasteur appears to have been motivated in this instance as in so many others by a strong belief in a hypothesis of his own making and his conviction was based on something other than a logical inference from the facts. The situation is parallel to Lavoisier's reasoning after he had been impressed with the results of his study of burning phosphorus. In the discussion of the atomic theory (p. 196) I have already taken issue with those who say that scientists approach a problem without preconceived ideas or without prejudice, and here is another example of the importance of a strong belief derived from but little evidence.

At the time Pasteur turned his attention to fermentation, a considerable amount of work had been recently devoted to alcoholic fermentation. It had been recognized that a living organism, yeast, was always associated with the fermentation process, the conversion of sugar into alcohol and carbon dioxide. But the predominance of scientific opinion was to the effect that the formation of alcohol was a consequence of the decomposition of the dead yeast cells. This was essentially the view of the great German chemist Liebig. He postulated that some sort of a sympathetic vibration was transmitted to the sugar molecule by the decomposition of some complex material in decaying vegetable matter. The issue between him and Pasteur was clear from the day Pasteur started to study fermentation until Liebig's death in 1875. Pasteur regarded the life process of the yeast as the cause of the change of the sugar to alcohol. In effect he said: Without life, no fermentation. This may be regarded as a second grand hypothesis of Pasteur. Liebig regarded the life process as irrelevant; the important points were the materials in the yeast. He facetiously compared people who regarded the living organisms as important with the man who imagined the Rhine to be driven by the row of water mills which he saw across the river at Mayence.

The issue is of importance not only because the controversy between these two giants of the mid-nineteenth century was productive of much fine research but because it illustrates the difficulties of defining terms in biology.

After all, what was meant by the word fermentation? The production of alcohol from sugar in the presence of yeast, of course. This was the ancient example. But how about the production of lactic acid (and some amyl alcohol) from sugar solutions? This seemed to occur spontaneously at certain temperatures particularly in the presence of chalk and albuminous material. Was this fermentation? It was so designated when Pasteur set out to study it; and as a result of this study Pasteur demonstrated the presence of a microscopic living organism and isolated the "lactic acid yeast" (actually a bacterium). How about the processes Liebig himself had studied as a young man almost a generation earlier? For example, the change which took place when bitter almonds were crushed with water; this was a chemical reaction involving the formation of an oil (bitter almond oil) from a water-soluble material. Liebig showed that the agent for this process was something in the skin of the bitter almond. If this reaction were to be included under the term fermentation, Pasteur was in trouble from the beginning, and to the end of his life as well. For there was no living microorganism present in this case. Pasteur, as far as I am aware, never even tried to find the equivalent of the yeast or lactic acid ferment. Rather he dismissed this and similar cases as not being "true fermentation."

In one of his earlier papers on this subject Pasteur reports the results of his studies in these words: "I found that all fermentations *properly so called*, lactic, butyric, the fermentations of tartaric acid, of malic acid, of urea, were always connected with the presence and *multiplication* of living organisms. According to my views, albuminous materials were never ferments but the food of ferments. True ferments are organized entities." By these last two words

Pasteur means living organisms. The italics are mine, not Pasteur's, but they bring out the unconscious assumption involved in Pasteur's second grand generalization. And it is true that he could buttress his exclusion of some bio-chemical changes from his definition of fermentation be-cause they involved less drastic molecular changes than did the others. Organic chemistry, however, had not suffi-ciently developed when Pasteur was first engaged in this work to make the distinction as clearly as we might today.

Pasteur had indeed demonstrated that for the *changes he listed* the multiplication of living organisms was essen-tial; not only the presence but the growth of the organisms was necessary. The difference is all-important. For the dead cells, according to Liebig and his followers, could yield the decomposing materials necessary to cause fermentation. But Pasteur, by the most brilliant use of the microscope and by the invention of many techniques which became stand-ard in microbiology, proved up to the hilt that the chemical changes in the cases mentioned went hand in hand with the growth of the microorganisms *in the absence* of air. Indeed, before long he advanced still another grand hy-pothesis which he expressed in the words, "Fermentation is life without oxygen."

Let us now consider what subsequent generations of investigators discovered and how matters stand today. At the close of the nineteenth century a German scientist dis-covered that under high pressure a juice could be pressed out of a mass of living yeast which contained "something" that brought about alcoholic fermentation. This experiment if it could have been performed by Liebig would have been powerful ammunition in his hands. For there is no doubt about it, the formation of alcohol from sugar can be accomplished by a product of a living cell without the living cell being present.

A vast amount of work by biologists, biochemists, and chemists in the last fifty years enables us to construct a conceptual scheme in which the Liebig-Pasteur controversy finds a ready place. The changes which occur in the life process are, we now believe, all specifically catalyzed reactions; that is to say, transformations which occur at an appreciable rate only in the presence of minute amounts of special substances which we call *catalysts*. The catalysts which occur in nature we call *enzymes*. All of them appear to be proteins and many have now been isolated in pure crystalline form. The situation in brief is as follows: the formation of bitter almond oil (Liebig's first observation) is an example of a change brought about by an enzyme which is easily removed from living or dead cells; hence it is no trick to bring about such changes without the presence of living organisms. The change from sugar to alcohol, or sugar to lactic acid, however, and all of Pasteur's other "true fermentations" are brought about by enzymes which under usual conditions do not leave the living cell. Therefore, for these changes to occur the cell must be alive and vigorous, for only under these conditions will the sugar penetrate the living cell and be transformed by the *intracellular* enzymes into the products (alcohol or lactic acid) which leak back into the solution.

Who was right—Pasteur or Liebig? I should say neither. Liebig's ideas about the sympathetic vibration of decomposing material were not fruitful and will not accommodate the observations. Furthermore, his failure to recognize that Pasteur's generalization might have at least a limited validity was certainly scientific blindness. Pasteur's generalization was wrong but it was marvelously fruitful. Perhaps as a cautious skeptic talking in terms of fruitful concepts or conceptual schemes I should refuse to answer my own questions. The reader may note that the biological influence got the better of me in the opening sentence of this paragraph and I asked a common-sense question. But the inability of anyone to give a clear-cut answer I may claim as evidence on the side of a skeptical chemist. However, I resist the temptation to press the issue further.

Let me conclude by emphasizing the difficulty of defining the term fermentation on the one hand and establishing a correlation between a living organism and a complicated series of events on the other. Substitute for the word fermentation, pneumonia or typhoid or measles or typhus, and for ferment, bacteria or agent and you meet the problems of disease to which Pasteur made so many contributions. It would be highly rewarding to review some of the significant episodes in the development of the "germ theory" of disease and then pass on to the work on viruses in this century. But rather than attempt this excursion into the field of medicine I refer the reader once again to the biography of Pasteur by Dubos. There one can see both the similarities and the differences between the methods of the applied biologist and those of the chemist or the physicist. In addition, the close relation between pure and applied science in the biological field to which I earlier referred is made manifest by the story of Pasteur's life. The work of this investigator is a rich mine for those who are curious about the methods of the experimental sciences. Indeed, one of Pasteur's earliest controversies at a time when he was just entering the biological field is so instructive that it requires a separate chapter. From physiology and biochemistry we turn to the controversy concerning the spontaneous generation of plants and animals.

CHAPTER NINE

Experiment and Observation
in Biology: Illustrations from
the Controversy concerning
Spontaneous Generation

T HE CONCLUDING sections of the preceding chapter
dealt with Pasteur's study of fermentation. Some insight was thus afforded into the difficulties of defining and
studying the life processes; an introduction was provided
to those two branches of biology which are today designated as physiology and biochemistry. In this chapter I
wish to illustrate the methods of experimental biology by
considering investigations in which biochemistry is but
little involved. In particular I shall call attention to the
difficulties of controlling the variables in biological experimentation and the need for a type of procedure for which
the phrase "control experiment" has been coined. To this
end a study of different episodes in the controversies concerning the doctrine of spontaneous generation is rewarding. We shall review very briefly the work of an Italian in-

vestigator in the seventeenth century, a controversy between an Englishman and an Italian in the eighteenth century, and a vigorous controversy involving French and English scientists in the third quarter of the nineteenth century. In this sampling of the work of biologists we shall see that there is no sharp line which separates observation from experiment, though in the preceding chapter we found it useful to make a distinction between systematists and experimentalists. Whitehead long ago pointed out that there was no real methodological distinction between astronomy and physics though the former may be regarded as an observational and the latter as an experimental science.

The relations between observation and experiment are very simply illustrated by Francisco Redi's study of the alleged spontaneous generation of worms in putrefying flesh. As I have already indicated, this early investigator combined the habits of the naturalist and physician with the experimental methods of the Accademia del Cimento. In 1668 Redi published the results of his studies of the formation of worms in meat. He demonstrated quite convincingly that contrary to opinions previously held the worms which appeared in meat after several days were *not* formed spontaneously. On the contrary they originated from eggs deposited by flies. In his account Redi starts with a description of the careful observations of a natural phenomenon under usual conditions. He clearly indicates how he proceeded from observation to experiment.

Redi first recounts what he observed on the surface of the meat kept in an open box for many days in Florence in mid-July. Not only worms but also small objects he calls eggs (actually they were pupae) appeared on the surface of the meat. He also noted the hatching of many flies. Of these Redi said: "There were to be seen many broods of small black flies . . . and almost always I saw that the decaying flesh . . . was covered not alone with worms but with the eggs from which, as I have said, the worms were hatched. *These eggs made me think of those deposits dropped by flies on meats, that eventually became worms,* a fact noted by the compilers of the dictionary of our Academy, and also well known to hunters and to butchers who protect their meats in summer from the filth by covering them with white cloths."

Here is the record of the naturalist, the fieldworker, the careful observer of biological processes as they naturally occur. These observations as emphasized by the words I have italicized in the quotation seem to have led to the hypothesis that all the worms originated in the deposits of flies. For Redi in the next paragraph notes, "Having considered these things, I began to believe that all worms found in meat were derived from the droppings of flies and not from the putrefaction of the meat." Here is an example of the grand working hypothesis in a biological field. From it consequences follow which can be tested by specific experiment, that is, by observation of some artificial situation. Redi was clearly following the pattern of the members of the Accademia del Cimento who were at this time studying pneumatics and hydrostatics. The experiments which he performed were quite simple but he felt them to be essential, for he writes, "Belief would be in vain without the confirmation of experiment."

So Redi, to test his hypothesis that flies are essential for the production of worms in putrefying flesh, proceeds to eliminate the flies. To this end he seals up the meat in glass flasks and notes with satisfaction that even "though many days had passed" no worm was seen. But in the same paragraph he tells of another highly significant observation, namely that samples of the same meat placed in similar *open* flasks at the same time soon "became wormy and flies were seen." Here is a good example of a recurring phenomenon in experimental biology, the *control experiment.* I shall have more to say about it after we complete the summary of Redi's work.

In keeping out the flies Redi likewise had prevented the air from circulating. One might say it was this rather than the absence of the flies that was responsible for the absence of worms. To test this point Redi used a simple procedure; he closed the flasks with a "fine Naples veil that allowed the air to enter." Again he observed no worms. With this evidence before him he considered the problem solved, and as far as I am aware, this particular case of alleged spontaneous generation was never reopened. Yet it is worth taking a moment to see just how far a skeptic might push his doubts as to what conclusions can be drawn. In so doing some similarities and some differences between biological and physical experimentation may be brought to light.

THE CONTROL EXPERIMENT

The important principle we have met before, the control of variables, appears in these experiments but in a somewhat altered form. One may analyze Redi's procedures and say that he in effect recognized three variables: (1) the flies, (2) the circulating air, (3) a sum total of effects such as time, place, warmth, kind of meat. The methods of testing the effect of the first two variables are as clear a case of common-sense experimentation as could possibly be found and in no way differ from the procedures of Robert Boyle. The third variable is the peculiar one. Redi makes no specific mention of it but it is clearly implied by the circumstance of his "control experiment." By placing side by side open and closed flasks containing samples of the *same* meat, Redi in effect answered any critic who might say, "but perhaps the meat wouldn't have produced worms on that particular day even if it had not been sealed up."

The essence of the control experiment in biology is an attempt to insure that only the variable being tested is affecting the results. The method is not confined to the biological sciences. Perier (p. 75) was using a control experiment when he had an observer watch a second barometer at the foot of the Puy-de-Dôme. But the method has special importance in biological experimentation because so often a host of variables of unknown nature may be involved. One endeavors to eliminate them by running in parallel two or more experiments identical in every respect except one, so that whatever differences are observed will be due only to the single, known variable.

There are other differences between the techniques of the biological and the physical sciences which are revealed by an examination of Redi's experiments and his conclusions. It is often hard to decide how wide a generalization may be made on the basis of the experimental facts. Indeed, what were the "facts" in the restricted sense in which we are using the word in Redi's case? The reproducible experimental situation would seem to be: "If flies have no access to the meat, there are no worms." But this has only been shown to be true in Florence in mid-July. Will it be true in all parts of the world and for all kinds of meat? What do you mean by the words "meat" and "flies"? Something no vaguer in 1668, to be sure, than what was implied by the words "fire," "oil," and "sulfur" as then used, but far vaguer than the words "red oxide of mercury" as used by

Lavoisier and Priestley. In short, to assure reproducible material, to define conditions accurately, is a matter of great difficulty in the field of experimental biology. Although the same problem arises throughout all the sciences, the degree of difficulty is certainly greater here than in physics or chemistry. How Pasteur handled this problem and how even he stumbled we shall soon see.

It would clearly be unsafe to conclude that spontaneous generation never occurs, even if experiments with all kinds of meats in all manner of climates showed there were worms only when flies had access. Indeed, Redi himself believed that worms spontaneously appeared in plant galls.

Jumping to the mid-twentieth century, it seems quite clear that there is no way of testing and therefore no way of disproving some such statement as "Somewhere on the earth's surface living organisms are today being formed from nonliving matter" (though I doubt if anyone at all versed in biology believes such a statement). On the other hand, if someone were to come forward as people have from Redi's time almost to the present day and declare, "Under such-and-such conditions spontaneous generation occurs," then this becomes a testable proposition, although it is by no means a simple matter on which to obtain conclusive evidence, as the controversies of the later nineteenth century prove.

Someone may object to my using the words "conclusive evidence," for up to this point I have attempted to speak cautiously in terms of experiments which are in accord with or not in accord with conceptual schemes. And this calculated slip of the pen illustrates how much nearer even nineteenth-century biology is to common sense than is nineteenth-century physics or chemistry. Even the most cautious skeptic can hardly fail to fall into the habit of talking about cause and effect when reviewing the work on spontaneous generation.

Clearly one is here very close to the type of common-sense experimentation involved in everyday life, as when we try to discover the cause of a disturbing noise in the night. We might try opening this door or moving that shade or closing this window till the noise stops. Only a philosopher in his most philosophic moments would speak of our midnight hypothesis which led to these experiments as a concept or conceptual scheme! The same impatience may overcome the reader when I point out that the difference between Torricelli's concept of "a sea of air" and Redi's that "flies drop eggs that hatch out worms on putrefying meat" is only a question of degree. But the difference of degree admittedly becomes almost one of kind. Contrast, for example, the abstract modern concepts of electrons, protons, and neutrons with the equally modern ideas of "rickettsia" (very small organisms) as the *cause* of spotted fever. In physics we seem to be dealing with hypothetical entities far removed from our sense impressions; in discussing the agents which "cause" disease we *seem* to be talking about tangible things. But the closeness to common sense in the latter case is more apparent than real. If one attempts to define the concept of spotted fever or characterize the species of rickettsia "responsible" for spotted fever, a great deal of trouble develops; one must bring in a mass of observations, some highly empirical, some far removed from common-sense ideas. Difficulties are at hand similar to those referred to in connection with defining fermentation.

CAUSE AND EFFECT IN BIOLOGY

In medicine and experimental biology we are dealing with a continuous transition between the concepts of common sense and of science. To the degree that one has a "feel" as to the "reality" of a concept one tends to talk about cause and effect and introduce such firm words as "conclusive evidence." To the extent that the ideas are new or foreign we more readily speak of concepts and conceptual schemes.

The ready applicability of the common-sense ideas of cause and effect to biological phenomena does not depend solely on an instinctive belief in the "reality" of biological concepts. One highly important difference between biology on the one hand and physics and chemistry on the other turns on a difference in the ordering of events in time. If one event follows another, we are ready to consider that the first *may be* the cause of the second; but not vice versa. If repeated observation shows that event A always precedes event B, we accept as a matter of common sense that A is the cause, B the effect, although we realize that there is possible a long argument as to whether some earlier event was not the "real" cause of B, or perhaps of both A and B. A boy throws a stone through a neighbor's window. What is the cause of the broken glass? The stone, the boy, the friend who put him up to the defiant act? The important point is the sequence of events in time. Except in a moving picture run backward we do not observe in common life such a series of events as broken window, unbroken window, stone near the window, stone in boy's hand, etc.

Biological phenomena are events in time not unlike the simple case just cited. Not only the naturalist but the casual observer notes such sequences as bud, flower, fruit, and never the reverse. Even the experimental biologist under the most artificial conditions cannot do the equivalent of "running the film backward"; he must accept as a "fact" that such events as egg, chick, hen or rooster occur in the order named. If, therefore, flies are seen to drop "eggs" on meat and worms issue from the eggs, we say the flies are the "cause" of the worms. In more complicated cases we speak of hunting for the cause of this or that biological phenomenon.

Contrast the situations just considered with the simple chemical experiment of Lavoisier and Priestley with red oxide of mercury. Heating metallic mercury for a considerable period of time just below the boiling point in air results in the formation of the red oxide (p. 189). The red oxide heated to a still higher temperature yields back the mercury and the oxygen gas. Liquid mercury and oxygen may *precede or follow* red oxide of mercury in the course of time, depending on the temperature. Which is cause and which effect? Or turn back to a consideration of the elementary principles of hydrostatics (p. 129). Water seeks its own level as demonstrated by pouring water into a two-armed vessel. By blowing into one side or the other the level in the right or left hand may be slightly raised, but when the air pressure is equalized above both sides the levels are the same. One can easily contrive to have a time sequence in which either the level on the right or the level on the left is higher. In short, the process is reversible; so too, it may be noted, is the formation of red oxide of mercury provided one varies the temperature and allows ample time. In physical and chemical processes, then, the experimenter can frequently alter the order of events. It is obviously complicated in such instances to speak of cause and effect. This is one reason why this terminology is far less common in the physical than the biological sciences.

Another reason why cause and effect are of doubtful value in physics and chemistry even for pedagogic purposes is the difficulty of treating any one of several variables as *the* cause. For example, hydrogen gas burns in air to form water. What is the "cause" of the flame? The hydrogen, the oxygen, the heat evolved, the affinity of hydrogen atoms for oxygen atoms, the electronic configuration of the respective atoms? And this is a simple chemical reaction. If we turn back a few pages and re-examine Pasteur's study of fermentation we can see he was maintaining a thesis as to the cause of a complex reaction—fermentation—which he believed to be the presence of growing microorganisms. A reflection on the outcome of this story, however, provides a clue to the conditions under which the notion of cause and effect in science is useful and those when it is not. Under the heading "a growing microorganism," Pasteur actually summed up a host of variables—cell walls, enzyme systems, the reproductive mechanism of the organisms, etc. But because he could embrace a number of unknowns in a single package, he could for the time being study the relation of this package of ignorance to certain chemical changes. Under these rather primitive scientific circumstances, asking "What is the agent that causes sugar to yield lactic acid?" has experimental meaning. Today with our knowledge of enzymes one would have difficulty in answering the question, "What is the cause of fermentation?"

This digression about the words used in "explaining" science has perhaps served a purpose if it has alerted the reader to the complexity of variables in various fields of modern science. It may likewise suggest that some of the statements one hears about cause and effect having disappeared from modern science are to be listened to with grave suspicion. In this book I have not considered quantum phenomena and the much advertised "uncertainty principle" in modern physics. But perhaps the review of scientific methods and modes of reasoning has gone far enough to suggest that one must either be prepared for a lengthy and penetrating philosophic analysis of the theory of knowledge or else be content to use common-sense terms to the degree they seem convenient. As a framework of reference within which the scientist may plan experiments, the terms, cause and effect, may in some areas be useful, in others only sources of confusion.

THE EIGHTEENTH-CENTURY CONTROVERSY
CONCERNING HETEROGENESIS

To return to the discussion of the doctrine of spontaneous generation, we may note that as a consequence of Redi's experiments and similar observations, the idea of the spontaneous generation of common plants and animals seems to have been given up. But the discovery of the world of microscopic organisms by the microscopists of the late seventeenth century opened up a new area in which the controversy could flourish. The origin of the mass of minute organisms revealed by the microscope in all manner of animal and plant extracts and scrapings was a subject of debate among the biologists of the eighteenth century. One highly placed French naturalist, the Comte de Buffon, became a powerful protagonist of the doctrine of heterogenesis, as the notion of spontaneous generation was later designated. Buffon considered all living matter to consist of organized particles essentially indestructible but capable of entering into different combinations. These "organic molecules" constituted the essence of life. These ideas were put forward, it must be noted, before the chemical revolution (Chapter VII) and a half-century before Dalton's atomic theory (p. 196). Buffon strongly opposed those who maintained that microscopic organisms, like larger plants and animals, had specific living precursors. Microscopic germs—the equivalent of eggs or seeds—he thought to be nonexistent.

An English amateur biologist, John T. Needham, collaborated with Buffon in some of the latter's writings and supplied what he regarded as convincing experimental evidence as to the ability of dead material to *regenerate* living matter. He seems to have been the first to use elevated temperatures in an effort to kill or destroy all living organisms in a liquid or solid material. Thus he corked up "mutton gravy" in a glass phial and heated the phial in hot ashes. This procedure he claimed would kill all pre-existing germs. Yet in a few days the phial after cooling off swarmed with microscopic organisms. We would say today Needham's experimental ideas were good but his interpretation of the results was erroneous. Indeed, this was the opinion of a contemporary, the Italian naturalist Spallanzani. Like Needham he used elevated temperatures to destroy any "germs" present in the various mixtures of plant and animal tissues whose putrefaction and decay he wished to study. Infusions made by soaking a variety of seeds in warm water were his favorite objects of investigation. He concluded from his experiments that if one took adequate precautions and heated the infusions *long enough,* no living organisms would subsequently appear.

Some who have studied Spallanzani's papers feel that his experiments should have settled the matter. That he was ahead of his time is clear; for the upshot of the Needham-Spallanzani controversy was that both remained unconvinced and the scientific world continued to be divided. Pasteur, reviewing the work of these two opponents from the vantage point of nearly a hundred years later, points out clearly why the issue was not settled by Spallanzani's work. The point is of more than passing interest for it is another illustration of the difficulty of defining biological concepts in experimental terms. Needham's defense against Spallanzani boiled down to this: in order to prevent the subsequent growth of microorganisms in infusions of plant and animal material the Italian had subjected the materials to the temperature of boiling water for periods of time far longer than Needham believed were necessary in order to destroy living organisms; Spallanzani had "weakened or perhaps totally destroyed the vegetative force of the infusions"; he had tortured the material, the Englishman maintained. Needham and Buffon, we must remember, postulated the presence in dead animal and plant material of a "vital force" which was different from specific living "germs." The germs should be killed by short exposure to boiling water, for this exposure would cook an egg and "kill" small plants and animals, but the vital force could stand only short periods of "cooking"; it was far too sensitive to stand such prolonged boiling as Spallanzani used. Such was Needham's position.

If you define a vital force or even "organized molecules" in terms of resistance to elevated temperatures you end with a doctrine of heterogenesis that Spallanzani's experiments do not invalidate. While we may feel that the notion of a vital force represents prescientific thinking, the idea of sensitive organized molecules which are altered by exposure to boiling water is by no means foreign to the twentieth-century chemistry of proteins.

Not only was the controversy between Needham and

Spallanzani a draw as far as scientific opinion was concerned, but a remarkable experimental discovery in the
early nineteenth century still further complicated the issue.
Shortly after 1800 an enterprising French confectioner by
the name of Appert had applied the methods of Needham
and Spallanzani to the preservation of food. He was the
inventor of the process we now call "canning," for he
showed that if you fill a bottle practically full of food material, heat it in boiling water for some time, and stopper it
well while still hot, the material will keep for long periods
of time. Here, by the way, is another example of the success
of a highly empirical procedure, for when this scientific
discovery was taken over by the practical arts it was still
an open question what were in fact the variables which
were altered by the process we call "sterilization by heat."

Not only was what occurred in this sterilization by heat
an open question, but the rapidly developing new science
of chemistry yielded a false clue. A distinguished French
chemist soon submitted Appert's procedure to chemical
test and found that the air which remained above the
"canned" foodstuff *contained no oxygen*. He therefore concluded that the significant variable in the preservation or
putrefaction of animal or plant material was the presence
of oxygen. Oxygen, in other words, might be the vital
principle which Needham complained Spallanzani had destroyed.

One more bit of history before we come to the detailed
examination of Pasteur's classic work on spontaneous generation. In 1837 a German investigator introduced a new
experimental technique which was to play an important
role in later work on heterogenesis. He showed that air
which had been *heated* could be introduced into a flask
containing meat juice without causing the latter to putrefy.

These findings showed that the presence or absence of oxygen was not the significant variable. They indicated that it was the dust in the air which mattered, dust presumably carrying germs. This was rendered still more probable by the work of two other Germans twenty years later in which air filtered through cotton proved to be in general the equivalent of heated air in experiments with material capable of putrefaction. In retrospect the evidence seems convincing, for we have become accustomed to the view that "germs" must be present as a necessary condition for the appearance of microorganisms in foodstuffs or other plant and animal mixtures. The correlation between the presence of the microorganisms and the putrefaction or fermentation is, of course, the problem which led Pasteur from chemistry to biology (p. 223).

Pasteur's interest in spontaneous generation was a natural outcome of his study on fermentation. Indeed, in his first important paper on spontaneous generation he also discusses his views on fermentation. But as his biographers have made clear, Pasteur felt impelled to study the question because of a paper by Pouchet. This naturalist, the director of the Museum of Natural History in Rouen, had become convinced of the possibility of spontaneous generation. Pouchet's experiments in support of his views were published in 1858 and an elaborate reply by Pasteur in 1862. The controversy continued vigorously for the next few years, and then, Pasteur appearing to have won a complete victory, interest subsided. But in the seventies an English doctor, Henry C. Bastian, took up the cudgels for the doctrine of spontaneous generation and the subject was reopened with highly beneficial results for the progress of science. By the end of the eighties, however, the evidence against heterogenesis seemed overwhelming, and though

Bastian gave his views in a book published as late as 1910, few proponents of heterogenesis survived into the twentieth century.

Pasteur's paper of 1862 on "The Organized Corpuscles Which Exist in the Atmosphere" is one of the great documents of experimental science. In the historical introduction the author, after reviewing the work of Redi, Needham, Spallanzani, and the more recent experiments in Germany, writes as follows (the remarks in brackets are mine):

"When after the researches of which I have just spoken, a skillful naturalist of Rouen, Pouchet (corresponding member of the Academy of Sciences) announced to the academy the results on which he thought he could base in a definitive manner the principles of heterogeneity [spontaneous generation] no one could point out the true cause of error in his experiments. Soon the French academy, realizing how much remained to be done, offered a prize for a dissertation on the following subject: Attempts by well-conceived experiments to throw new light on the question of spontaneous generations.

"The problem then appeared so obscure that Biot [a distinguished French physicist], whose kindness with regard to my work has always been unfailing, expressed his regret at seeing me engaged on these researches. He exacted a promise from me to abandon the subject after a limited time if I had not overcome the difficulties which were then perplexing me. Dumas [the dean of French chemists], who has often joined with Biot in showing kindness to me, said about the same time, 'I should not advise anyone to spend too long on this subject.'

"What need had I to concern myself with it? Chemists ran into a collection of extraordinary phenomena twenty years before which are designated by the generic name *fermentations.* All require the concurrence of two substances: one known as *fermentable* material, such as sugar; the other nitrogenous material, always in the form of an albumin-like substance. The theory which was universally then accepted was as follows: the albuminous substances undergo a change on exposure to air (a special oxidation of unknown nature); this gives them the character of a *ferment,* that is to say, the property of subsequently acting, through contact, on fermentable substances."

Pasteur then discusses his work on the lactic acid fermentation which was considered in the preceding chapter (p. 226) and contrasts his conclusion with Liebig's ideas. He then continues:

"One knew that ferments originated from the contact of albuminous substances with oxygen gas. One of two things must be true, I said to myself: either ferments are organized entities and they are produced by oxygen alone, considered merely as oxygen, in contact with albuminous materials, in which case they are spontaneously generated; or if they are not of spontaneous origin, it is not oxygen alone as such that intervenes in their production, but the gas acts as a stimulant to a germ carried with it or existing in the nitrogenous or fermentable materials. At this point, to which my study of fermentation had brought me, I had to form an opinion on the question of spontaneous generation. I might perhaps find here a powerful weapon to support my ideas on those fermentations which are properly called fermentations.

"The researches which I am about to describe were consequently only a digression I was forced to make from my work on fermentations. It was thus that I was led to occupy myself with a subject which up till then had taxed only the skill and wisdom of naturalists."

Taken together with our knowledge of the factors which induced Pasteur to study fermentations (p. 224), we have thus a complete account of the transformation of a chemist into a biologist. But these paragraphs from Pasteur's paper are worth reading quite apart from the light they throw on the forces which move a genius. For Pasteur clearly states the problem of the relation between putrefaction and fermentation and the question of spontaneous generation. He had to determine for himself whether procedures like Appert's worked because the oxygen was used up or because the germs were destroyed by the boiling process. He proceeds to amass an amount of experimental evidence which, as compared to the work of his predecessors, was overwhelming.

I shall make no attempt to summarize even this one paper by Pasteur. However, I shall discuss the type of experiment he used and some of the difficulties of interpretation. We are here dealing with *converging evidence,* it will be noted, for no single set of experiments by themselves would appear sufficient to answer the objection of his opponents, the believers in spontaneous generation. Pasteur repeated and confirmed the work of his immediate predecessors in this field. He showed that air which had been passed through a red-hot tube (calcined air) could be introduced into a vessel (suitably sterilized by boiling) containing fermentable material without starting fermentation. On the other hand, when ordinary air was passed into a similar vessel containing the same material, fermentation soon started (note the control experiment). The vessels were placed in a warm closet in every case to hasten the fermentation. The fermentable material Pasteur employed was what he called "sugared yeast water." It was an extract of yeast, to which sugar was added; it contained no living yeast, but in addition to the added sugar, nitrogenous substances and mineral salts from the yeast were present. In other words, by purely empirical procedures Pasteur had prepared a good "nutrient" medium from yeast. This choice of one particular experimental material had important consequences, as will be noted shortly. Pure empiricism is an important part of almost every experimental procedure.

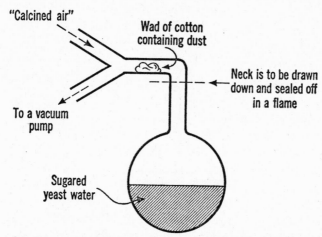

Fɪɢ. 31. Diagram illustrating Pasteur's technique. The sugared yeast water is boiled and the wad of cotton introduced into the tube. By repeated evacuation and filling with "calcined air," the ordinary air is replaced by calcined air. The flask is then tilted so that the cotton falls into the liquid and the flask is sealed off (Fig. 32).

Pasteur convinced himself that he could repeatedly prepare "sugared yeast water" under conditions which would prevent its subsequent fermentation in a warm closet. He then proceeded to use this knowledge as a basis for experiment. He collected the dust from ordinary air on a bit of cotton wool by sucking a considerable amount of air through a cotton wool filter. He then ingeniously contrived to introduce this bit of cotton into a sterile flask containing the sugared yeast water, under conditions where only *heated* air was likewise introduced (the experimental technique is illustrated by the simplified diagram shown in Figs. 31 and 32). In due course the flasks which contained the cotton showed signs of vigorous fermentation, while others into which cotton was not introduced did not.

The use of the "control experiment" is evident in this example of Pasteur's work. No one maintaining the negative

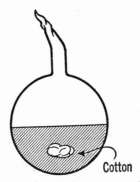

Cotton

FIG. 32. The flask of Fig. 31 after being sealed off.

position in this debate on spontaneous generation could do otherwise than declare that the failure of precautions to destroy germs or keep them out would result in fermentation. Therefore, it was essential to show that if everything else was the same *except* for the cotton, there would be no fermentation. Pasteur went further and performed a special experiment which was in the nature of a "control" on the others. He substituted asbestos for the cotton; the results were the same. Therefore, since two quite different filters behaved in the same manner, he argued that it could not be the introduction of the filter which started the fermentation. With the asbestos, however, he performed what seems to me his most striking experiment. He showed that an asbestos filter, filled with dust in the usual manner *but then heated,* when introduced into the flasks produced no fermentation; the same filter filled with dust but not heated caused fermentation to start when the experiment was performed in the usual manner.

Let us review these classic experiments of Pasteur for a moment in the light of what he wished to demonstrate. He believed that very small particles of living matter (germs) must be present before the fermentation of sugared yeast water could begin. These particles were too few and too small to detect *as such* even with a microscope. Therefore his demonstration had to be indirect. Nevertheless, he was able to show that a "something" in ordinary air had to be introduced into the flasks before fermentation commenced. Furthermore, this something could be collected with dust on a filter and was destroyed (or its effectiveness destroyed) by heating. What could a "something" thus defined in experimental terms be but a precursor of the microorganisms which grow in sugared yeast water under ordinary conditions? So far Pasteur was on safe ground; he never had cause to retreat an inch from the position taken as the result of these experiments. But his attempt to marshal still more evidence led him onto treacherous ground.

PASTEUR'S CONTROVERSY WITH POUCHET

To follow this part of the story, one must remember the view that had prevailed a few years earlier, namely that oxygen was the essential variable in putrefaction or fermentation. This was Pouchet's contention. Many experiments in which putrefaction or fermentation could be induced by the introduction of a small quantity of air could be accounted for on *either* the germ theory or the oxygen theory. The "calcined air" experiments of Pasteur were convincing, but he devised some additional and very simple experiments. He placed his sugared yeast water in a flask, boiled the solution, and then sealed off the neck of the flask in a flame. After cooling, the tip of the neck could be broken and a small amount of air then rushed in, as a partial vacuum had been formed in the flask when it was sealed at a higher temperature. Pasteur then sealed the tip again and placed the flask in a warm room. Now, he argued, if oxygen is the significant factor, *all* the flasks thus treated should behave in the same way, for they all had received a small inrush of oxygen when opened and sealed again. But many experiments showed that whether the material in the flasks fermented or not depended on *where* they had been opened and closed. Only rarely did *all* the flasks in a dozen or so treated in the same way show the growth of microorganisms. Indeed, when opened in the country, only 8 out of 73 flasks showed signs of fermentation in experiments performed as just described. And in a spectacular demonstration Pasteur showed that of 20 flasks opened and closed on a glacier, the Mer de Glace, only one subsequently showed the presence of microorganisms; while 10 out of 13 opened in a room in the inn at Chamonix showed signs of growth.

Dust was certainly less widely distributed in the air on a mountain than in a village inn; therefore it was reasonable to conclude that the striking difference reflected a difference in the distribution of the germs in the air. But the vital point was that air could enter many, many flasks and *not* cause fermentation. The analysis of the air in Appert's preserves had yielded a false scent. There is no oxygen in the air above canned foods. This is due to a slow secondary effect—the absorption of the oxygen by some material in the food. Such was Pasteur's explanation and it is the one we accept today. Pasteur summarizes his findings in his paper of 1862 by the statement: "It is not true that the smallest quantity of ordinary air is sufficient to produce in an infusion the organized life characteristic of that infusion."

Pouchet and his supporters were by no means convinced by Pasteur's paper. They proceeded to experiment on mountain tops themselves and obtained results in complete contradiction to those Pasteur reported. Bottles containing fermentable material were opened and closed on the summits of Mont Blanc and Monte Rosa and on a glacier in the Pyrenees. All Pasteur's precautions were taken, it was said, yet *in every case* growths appeared when the vessels were kept in a warm place. Pasteur naturally attributed these results to poor experimentation, for it must be emphasized again that every experimental error would seem to favor the heterogenists; every failure to destroy or exclude germs would, on Pasteur's hypothesis, yield results that *appeared* to indicate spontaneous generation.

A committee of the French academy was formed to settle the controversy between Pasteur and Pouchet. Pasteur produced his flasks which showed no signs of fermentation although they had been opened and closed. His evidence was most convincing. For reasons that are not clear Pouchet and his collaborators raised trivial objections about the conditions set down by the committee for the test, refused to carry out their experiments, and finally withdrew. The committee decided in favor of Pasteur. The victory seemed complete in 1865. But ten years later it became apparent even to Pasteur that Pouchet had withdrawn from the field of battle far too soon.

Not that any evidence for heterogenesis was found that could stand the test of rigorous experimentation. But what was discovered indicated that Pouchet's experimental results were by no means the consequences of faulty manipulation. The trouble lay elsewhere. Pouchet had used infusions of hay, Pasteur sugared yeast water. Both investigators and the scientific onlookers had *assumed* that the nature of the fermentable material was of no consequence. Actually, it was a variable of the first order of importance. Why? Because the microorganisms naturally present in hay form spores, a sort of resting stage in the life history of these bacteria. These spores are highly resistant to heat but do not give rise to a vigorous growth of microorganisms *except in the presence of oxygen.* Therefore, the boiling sufficient to sterilize sugared yeast water is quite inadequate for hay infusions. The introduction of oxygen does start up growth *in every case* in hay infusions which have been inadequately sterilized (as were Pouchet's). In short, with Pouchet's flasks, the presence or absence of oxygen was the important thing, not the presence or absence of germs in the air.

All these difficulties with sterilizing certain types of infusions came to light in the 1870's as a consequence of still further controversy. The protagonist for heterogenesis was the English physician Bastian, already mentioned. On the other side Pasteur had a firm and expert ally in the physicist John Tyndall. To go into the details would require another chapter, but, in a word, Bastian forced Pasteur and Tyndall to revise their ideas on what procedures were required to destroy all precursors of living organisms. Temperatures above the boiling point of water were sometimes necessary; from this time on, Papin's digestor (p. 107), now called an autoclave, became a feature of biological laboratories and, soon after, of hospitals as well.

One can almost hear the ghost of Needham raising an objection to the techniques of the bacteriologists as they were finally developed in the 1880's. If there were a sensitive vital principle in vegetable or animal matter, it was indeed tortured by heating to temperatures well above the boiling point of water and in some instances for long periods of time. But the day of such vague ideas as that of a vital principle was rapidly passing. Biochemistry and bacteriology were narrowing the definitions of their concepts. By the end of the nineteenth century the experiments of Pasteur and Tyndall had only historic interest insofar as the controversy about spontaneous generation was concerned. But their importance was recognized for other reasons; their significance lay in the fact that these experiments were the firm foundation for the techniques of the bacteriologist and microbiologist. Following the lead given by Pasteur in his first paper on lactic acid fermentation (p. 227), scientists had learned how to isolate and grow pure strains of microorganisms. Therefore, instead of a miscellaneous collection of microorganisms forming when dust was introduced into a nutrient material such as sugared yeast water, one could at will grow one organism or another by suitable inoculation. One could come very close to realizing the condition of seeing (with a microscope) the single germs which when introduced into a sterile medium were the starting point of the subsequent growth. In other words, the advancing techniques and concepts of bacteriology placed any proponent of spontaneous generation in the position of having to specify what organism he claimed could spontaneously arise. And it would be difficult for even Needham's ghost to claim that a sensitive vital principle could yield one kind of organism when sugared yeast water was inoculated with a drop of one fluid and another when the same medium was inoculated with a different fluid. The nitrogenous materials from the yeast, the meat, the hay, whatever is fermenting or putrefying can hardly be imagined to be other than "the food of ferments" and not the "ferment," to use Pasteur's own words.

The outcome of the work we have just reviewed seems so obvious that we accept the results today as part of common sense. By so doing we fail to realize the difficulties of relating concepts to experiment in experimental biology, and the significance of the story is thus often lost. It is well worth restudying the whole history of spontaneous generation in order to appreciate that, as in the case of fermentation, one is here viewing the transformation of vague common-sense notions into scientific concepts. And how tortuous may be the process of transformation! How difficult it is to reformulate common-sense notions carrying with them so much of a nonlogical background into ideas and terms related to experiment. The common-sense ideas we express by such words as "living organisms," "living precursors of microorganisms," or as "germs" have a psychological and sociological basis. They became concepts of science only slowly and as the result of a vast amount of work.

If we could follow the story into modern biology we would see some of these same troubles plaguing (but not blocking) the worker in pure and applied biology today. Is a "virus" alive or not? Is a disease agent not only a necessary but a sufficient "cause" of a disease? If not sufficient, what are the other variables? How do you define a disease? How specify a species of microorganism? How define a particular virus? To follow these and countless other questions would take us into the modern laboratories of departments of biology, agricultural stations, medical schools, hospitals, and special institutions. There we should see in progress investigations of problems Pasteur never even dreamed of, but we should certainly find methods and modes of thought essentially the same as those he employed. And if we were lucky enough to meet with a pioneer, a driving genius, we would probably find him as bold with his hypotheses on a grand scale and as certain of his preconceived ideas (one may even say his scientific prejudices) as was Pasteur himself.

But how about the origin of living organisms, some reader may inquire? If they don't originate spontaneously, how did all the vast number of plants and animals large and small ever get started? To answer this question or rather to face up to this question, one must consider our methods of studying what happened in the past. For while no one can categorically deny that spontaneous generation is occurring on the earth today, one can say that no phenomena have been studied which cannot be accommodated far better by the concept that for every living organism there is a living precursor. As to the investigation of the distant past, the following chapter will at least discuss some of the advances as well as the difficulties in that great scientific undertaking.

CHAPTER TEN

The Study of the Past

A RECENT WRITER on the growth of scientific ideas has
given it as his opinion that there have been three great
revolutions in modern science: the Copernican, the Newtonian, and the Darwinian. The reader will be aware that
up to this point at least I have made little or no reference
to any of these major events in the history of science. Nor
shall I do so in the remaining pages of this book. The reason
should be manifest. The present volume is concerned not
with the impact of scientific ideas on the intellectual climate
of the western world but rather with the methods by which
experimental science has advanced in the last three hundred years. Those who are concerned with the history of
scientific theories and their relation to changing opinions
about the origin of the world and its inhabitants are referred to such recent books as Wightman's *The Growth of
Scientific Ideas* and Butterfield's *The Origins of Modern
Science* as well as such classics as Whitehead's *Science and
the Modern World.*

Nevertheless, in the present chapter I shall come within sight, at least, of that ambiguous area bounded by theology, philosophy, and science. For I propose to consider briefly some of the special problems which arise when scientists and scholars consider the distant past; in particular I shall examine the methods and basic postulates of the sciences of geology and paleontology, with a side glance at some of the more general problems of cosmology. Now it is in exactly these areas of science that within the last one hundred years new ideas have developed which have had tremendous repercussions on the general outlook of the intelligent citizen of Christian nations. To be sure, there is today no novelty in evolutionary theories, and theologian and agnostic now quarrel most violently in their interpretations of the views of psychologists, anthropologists, and sociologists; yet it was only a little more than fifty years ago that Andrew White, the first president of Cornell, felt compelled to write his two-volume history of *The Warfare of Science and Theology.* As White states in his preface, he had been so harassed in the starting of a new university by objections from the more orthodox members of the Protestant churches that he was moved to strike back on behalf of freedom of scientific inquiry. His book is still well worth reading for many reasons. But I have referred to it here because even a cursory examination of the chapter headings shows that the warfare White describes has been for the most part a series of battles in regard to the interpretation of the past. Orthodox theologians were ranged on the one side, and on the other either the advocates of new scientific theories of the history of this planet or historians and scholars who were applying critical methods to ancient documents. Clearly a discussion of even the methods of students of the past brings one automatically within range of the weapons of highly partisan controversialists.

As long as one sticks to physics, chemistry, and experimental biology, a cautious approach to science such as I have employed will offend but few. For the reader may apply as much dogmatism as he likes to correct the picture and choose his dogma from whatever source he pleases. For example, those who insist on the reality of atoms, molecules, and genes will probably be only slightly annoyed by my refusal to grant these words any other meaning than names for conceptual schemes. They can readily accommodate their scientific faith to my skepticism. Such readers will at least agree that in order to understand the processes by which new ideas are evolved and tested by experiment it is advantageous to recapture the attitude of those who first regarded as tentative hypotheses, conceptual schemes we now take for granted. On the other hand, readers whose theology leads them to regard the scientific account of the world as quite incomplete may welcome my tentative approach and applaud my lack of scientific dogmatism. In short, the discussion of methodology presented in the preceding chapters may perhaps be found satisfactory (though incomplete) by pragmatists, logical empiricists, neo-Thomists, fundamentalists, Kantians, and agnostics. Perhaps only the militant believer in the complete adequacy of modern scientific explanations of the universe will be unduly irked by my unwillingness to be more dogmatic about the relation of scientific conceptual schemes to reality; but even he can perhaps be charitable in his condemnation, for the attitude which pervades the previous chapters may be regarded as a relatively harmless pedagogic device. The ardent believer in the doctrine of dialectical materialism, however, may well feel that this book is completely out of bounds. In so doing he would be quite consistent, particularly if he was committed to the doctrine now official in the Soviet Union. For it is important to remember that one of Lenin's significant philosophical treatises, written in 1909, was an attack against Mach's analysis of the concepts of physics. Mach's trenchant criticisms of some of the then current assumptions of science were held to be reactionary because they seemed to open the road for either skepticism or idealism. If one is going to found a philosophy on the findings of nineteenth-century science, then there can be no toleration of any doubts as to the finality and validity of these findings.

The approach to experimental science characteristic of this volume may well be compatible with a variety of philosophic and religious convictions. But if we are to be wary of dogmatism in science, we must be no less cautious as to philosophy, theology, and history. The acid of skepticism must be applied with equal boldness to religious documents and to scientific theories. Therefore, this chapter may offend both the traditionalist in religion and the dogmatic naturalist or materialist. To others it will be disappointing, for it will suggest too restricted a basis on which an individual may build his personal philosophy of life. It has been recently said that "the peculiar problem of the age lying ahead of us will be to reconcile science and wisdom in a vital spiritual harmony." To this most of us would agree, but we might well differ profoundly as to the nature of the obstacles to a reconciliation. They do not lie in the fields of physics, chemistry, or experimental biology; that much seems certain. The difficulties concern the record of the past and the status of certain documents. The traditional view of many theologians must be reconciled with the conclusions of critical biblical scholars and historians of religions.

A reconciliation must likewise be accomplished between orthodox theology and the biological sciences which deal either with the distant past or with the behavior of man as an individual.

All these difficulties may be overcome, I am convinced, but not on the terms frequently offered by the more orthodox members of certain of the Christian churches. They can be overcome, it seems to me, only if all the documentary evidence in support of the doctrines of Christianity or Judaism or of any other religion is subject to the same critical examination as the boldest would apply to the scientific explanations of man's origin and development. But to explore this subject further would require a digression far beyond the province of this book. Suffice it to say that an attempt to suspend final judgment as to the validity of the current dogmas of both science and theology would seem an essential element in a successful reconciliation. Skepticism without sneers may provide a cautious approach for men of many minds to both science and religion. And such an approach does not exclude the possibility of a variety of faiths, of diverse modes of expressing that spiritual wisdom which has come to the inhabitants of the western world through more than one historical channel.

The study of the past not only brings one into territory which has been fought over by theologians, scholars, and scientists but raises some difficult questions for all who insist on treating scientific theories as fruitful conceptual schemes. We have already seen that many scientific ideas have become so deeply embedded in our everyday view of the world that we find it difficult to draw the line between conceptual schemes and matters of fact. What were once working hypotheses on a grand scale and later became new conceptual schemes are now almost universally accepted as being descriptions of reality. In unguarded moments everyone will say that it is a fact that "we live on a globe surrounded by a sea of air," that "the earth goes around the sun," that "matter is composed of atoms," that "living organisms today originate only from living precursors." Yet, to understand science I believe it is important to distinguish the ideas thus expressed from such facts as "a suction pump will raise water not more than 34 feet at sea level and less at higher altitudes," and "red oxide of mercury yields mercury and oxygen when heated." But can we apply such a skeptical attitude to the conclusions of the students of the past? Can the results of geological and paleontological inquiries be handled with equal caution? And if so, what about the conclusions of the historians and the archaeologists?

Let us start with a segment of accumulative knowledge (p. 37) which falls outside our definition of science—for instance, by an examination of our knowledge of human events which we feel quite certain have occurred sometime in the last few thousand years. When we are talking about recorded history, common sense tells us we are talking about facts. Yet historic events appear to differ in certain respects from what we have hitherto cautiously regarded as facts. They differ from the facts of the physicists, chemists, and biologists much as the statement, "I spent the summer in Duxbury, Massachusetts, five years ago" differs from the assertion, "There is a pump in the kitchen which draws up water from the well."

At first sight there perhaps appears no difference between the two statements. Both, you may say, can readily be proved to be either true or false. But is there not a significant difference in the procedures involved in the verifications? The second assertion suggests at once a line of action

for anyone who doubts the validity of what has been as-
serted. "Go out in the kitchen and try the pump yourself,"
is the answer to the doubter. The statement of fact in this
case is in the nature of a prescription for an operation which
can be performed in the future by any number of unspeci-
fied individuals. Such are most of the factual statements
which constitute the body of practical knowledge and the
fabric of the natural sciences. For example, it is a fact of
chemistry that red oxide of mercury yields mercury and
oxygen in certain proportions when heated at a specified
temperature. Such a definition of a chemical compound is
clearly something fairly close to a set of rules which can
be followed by anyone who cares to do so any time in the
future.

The assertion that "I spent the summer in Duxbury five
years ago" carries conviction to the speaker because of his
firm belief in the infallibility of his memory. Yet to convince
a doubting Thomas requires something more complicated
and uncertain than saying, "Go do so-and-so and you'll be
convinced yourself." We all know that there are many true
statements about our past which we could not substantiate
with evidence. There are many others easily confirmed by
the testimony of specified individuals and the reading of
certain documents. Yet there may easily arise instances
where one is uncertain about the correctness of one's own
memory; in such cases, the process of verification is similar
to that employed to convince an investigating committee
or a jury of the truth of some fact of which we do not have
the slightest doubt. But no simple prescriptions for action
will suffice. Rather, there must be a mass of converging
evidence of the type so familiar to lawyers and historians.

There are tentative statements about the past which may
on examination prove to be true, false, or uncertain. The

skeptic will require more evidence perhaps than one who is ready to be convinced. But there is no question of the reality at least of the immediate past. We all have confidence in the general correctness of our memory. From a common-sense point of view we know that an assertion about some part of an experience must be either true or false. We instinctively feel as though we could be transported back to the scene in question. For our memory enables us to turn back the clock as far as the major events of our own life are concerned.

In discussing the operations of experimental science I have used the phrase "limited working hypothesis" to describe tentative statements which when verified we call facts (p. 51). Thus, if I find a bottle with a red powder I may construct the limited working hypothesis that the material is red oxide of mercury. (The procedure for testing this particular hypothesis has already been described.) Such limited working hypotheses are quite distinct, to my mind, from those hypotheses on a grand scale which often eventuate in conceptual schemes. If this has not been made clear in the earlier pages of the book, my attempt to convey an understanding of science has failed utterly. Now, a statement about an event in the immediate past would seem to partake of the nature of a "limited working hypothesis," though the procedures for verification, as we have seen, are somewhat different from those employed in the experimental sciences. A collection of limited working hypotheses does not constitute a working hypothesis on a grand scale, much less a conceptual scheme. Such an assemblage of tentative statements when confirmed represents merely a segment of empirical knowledge. Conceptual schemes are required to transmute such a mass of facts into a science.

Historical knowledge requires no conceptual schemes if the aim of the historian is simply to portray events as they really happened. A skeptic may be dissatisfied with the sufficiency of the evidence to support this or that alleged historical fact, but there is no question here of the reality of any conceptual scheme. (I am leaving aside the question of patterns in history, the interpretation of "historical forces," and problems raised by those concerned with philosophy and history.) No one in his right mind doubts that men and women very like ourselves were living in Rome, let us say, two thousand years ago. History as it is usually written engenders no debates such as we have previously considered in discussing such concepts as caloric fluid, atoms, and molecules. That is why I feel that there is a distinction of importance between history and science.

Those who disagree with my cautious approach to science may be quick to say that there should be no such arguments in science, either. For them the reality of scientific conceptual schemes as finally evolved is on the same footing as the reality of historical events. The significance of this position will be more clearly evident in a moment when we consider the sciences of geology and paleontology. Yet even those who define science as a quest for an understanding of reality would have to admit that there have been long periods when conceptual schemes now labeled "erroneous" were fruitful of new experiments and observations. And one can ask, "Where is the equivalent in the history of historical investigations?" It might be argued that a tentative historical reconstruction of some period was parallel to such a conceptual scheme as the caloric fluid or the luminiferous ether (both now obsolete except for pedagogic purposes). Admittedly, a professional historian in his most professional moments might declare that the significance of some recent historical findings lay in their fruit-fulness in stimulating further historical researches. But for laymen and many scholars, what the historian writes is of importance because we feel it to be a true description of past reality; otherwise we might content ourselves with fiction. We readily imagine that we ourselves could be by the side of Caesar when he crossed the Rubicon. Indeed, it would be my contention that the major objective of history is to increase our knowledge of the ways human beings behave under a variety of circumstances. The famous phrase of the seventeenth-century scholar John Selden sums up why the study of history should be central in all educational undertakings: "A study of the past may so accumulate years to us as if we had lived even from the beginning of time."

If the foregoing analysis be at all correct, quarrels be-tween dogmatist and skeptic are of a different character in the field of history from what they are in science. That faulty conclusions are possible in sifting historical evidence is all too plain. The tendency of human beings to falsify the record or to misinterpret the past is an everyday phenome-non. Furthermore, we are all aware how rapidly what seemed clear becomes obscure with the passage of time. As we push our historical inquiries back through the centuries, our knowledge becomes in general increasingly dim and uncertain. (And I am here using "knowledge" in the sense that I feel I have certain knowledge of what has transpired, with my participation, in the room in which I now write during the last few minutes.)

Someone has compared our knowledge of the past with that of a distant area. The analogy I think is not without its value. One can imagine the existence of an island in a lake just within sight of the shore but inaccessible. An observer on the lake shore can make out perhaps a few of the general topographical features or possibly see dimly a moving form which might be one animal or another. If a person had no method of approaching nearer to this island but could only move around the edge of the lake and make sketches from various angles, it might nevertheless be possible to construct a rough map of the land under observation. For a long while this might be the best one could do. But later, if armed with a telescope, the observer could greatly augment his information. And if he could fly over the island in a plane he might make a very good map without ever setting foot on it. So, too, converging evidence supplied by documents and monuments uncovered by the archaeologist has enabled each generation of historians in the last two hundred years or so to provide increasingly accurate pictures of many periods. As of today, the degree of certainty of the many reconstructions of past epochs differs enormously depending on the amount and kind of evidence available at present. Historians writing for other historians leave no doubt on this score; they indicate the sources of their information and estimate the correctness, more or less, of the statements made. But in history as in science, textbooks and popular accounts tell relatively little about how the conclusions presented were obtained; nor do they indicate in most cases the degree of uncertainty. Dogmatism all too readily gains the upper hand.

I recognize that it is far easier to find fault than to suggest remedies; a mass of footnotes and scholarly paraphernalia is discouraging to the student and the general public. Yet one may be permitted, perhaps, to register certain regrets at the inadequacy of the critical spirit in some popular historical writing. One can doubt if we have much clear knowledge, for example, of the Athens of Socrates or the Rome of Julius Caesar; yet the distinction between what is fairly certain and what is highly conjectural in a picture of the ancient world is rarely pointed out to the general reader. When one turns to ecclesiastical history, the quarrel of the skeptical layman with a historian committed to orthodoxy may be violent indeed. A scholar reviewing the work of other scholars in the field of biblical criticism has written, "History is not a deductive science and there are no rules for detecting fact. There are rules for detecting fiction, but that is a different thing altogether. It is not surprising, therefore, that very different views of the value of the Gospel according to St. Mark as an historical document have been put forward in recent times." One could wish that not only in writing religious history but in other attempts to depict events of several thousand years ago the author would indicate the divergences of scholarly opinion and give the reader some idea of the probable accuracy of his account of a distant scene.

CONCERNING THE AIMS OF GEOLOGY

In a volume about science, however, one is not entitled to devote too much space to other areas of accumulative knowledge. Having indicated how the skeptic may be prepared to suspect historical writing (particularly partisan accounts charged with emotional overtones), I should like to explore the assumptions and the methods of the geologist and the paleontologist. The history of the science of geology seems to indicate to one who is not a geologist that two different motives have activated the students of the earth's crust. On the one hand, a desire to reconstruct what actually happened in the extremely distant past is manifest; this seems the equivalent of treating geology as an extension of history, although the time intervals are vast and the uncertainties of the picture correspondingly very great. On the other hand, there is the desire to classify and correlate present-day observations and develop theories which will guide the geologists to make further observations. Here one seems to approach systematic biology very closely, except that the concepts necessary for the classifications involve a time scale of long duration. Are the theories of the geologist to be regarded as the equivalent of historical reconstructions with a high degree of uncertainty, or are they best regarded as conceptual schemes to be judged by their effect on the development of geology as a scientific activity?

There is little question that as geology is usually presented to the lay reader or the elementary student the subject is treated as though it were on a footing with history. But at the risk of incurring the wrath of my geological friends, I raise the question as to whether the equating of geology with earth history does not mislead the layman. The study of any period of human history of which our knowledge was as tentative as are the theories of geology would never be accounted of much value in terms of common sense. There would be no "accumulation of years to us" if our knowledge of historic events was as full of assumptions and uncertainties as those which characterize all working hypotheses on a grand scale in science. Geology seems to me to be far more closely related to biology than to history and increasingly similar in its methods to physics and chemistry. Indeed, one might take a number of the important generalizations which have developed in the geological sciences since 1800 and consider them much as we have analyzed the concept of the atmosphere, the phlogiston theory, the caloric fluid, and the atomic theory.

The conceptual schemes of the geologist have changed and evolved over the course of a century and a half as have those of the experimental scientist, but without them there would have been only a mass of unrelated empirical data; as to their fruitfulness, there can be no question. The conceptual schemes of the geologist like those of the physicists, chemists, and biologists have been fruitful not only as regards new scientific observations but in reducing the degree of empiricism in such practical matters as locating precious minerals, coal, and petroleum. In this century the fruitfulness of geological theories can be measured in terms not only of what may be observed in the field but also of what experiments may be performed in the laboratory. And many of the field observations are now essentially physical measurements such as finding the anomalies of the gravitational constant at various localities and recording the speed of seismic waves through the upper layers of the earth's surface. The origin of the concepts and conceptual schemes of modern geology is similar to that of the other natural sciences—speculative ideas, deductive reasoning, working hypotheses on a grand scale, and observation are all joined together.

Karl von Zittel in his *History of Geology and Paleontology*, written at the beginning of this century, speaks of the "heroic age of geology" which he identifies as the period from 1790 to 1820. He states that the characteristic feature of this age was a determined effort to discountenance speculation and to seek untiringly in the field and in the laboratory after new observations, new truths. This new outlook had a rejuvenating significance in the development of geology. Sir Charles Lyell in his famous *Principles of Geology* places the critical period in the evolution of modern geology slightly later. He speaks of the significance of the founding of the Geological Society and uses the following words in the edition of 1873 to indicate that same distrust of speculation to which the German geologist had referred:

"The contention of the rival factions of the Vulcanists and Neptunists had been carried to such a height that these names had become terms of reproach; and the two parties had been less occupied in searching for truth than for such arguments as might strengthen their own cause or serve to annoy their antagonists. A new school at last arose which professed the strictest neutrality and the utmost indifference to the systems . . . and which resolved diligently to devote its labors to observation. The reaction provoked by the intemperance of the conflicting parties now produced a tendency to extreme caution. . . .

"But although the reluctance to theorize was carried somewhat to excess, no measure could be more salutary at such a moment than a suspension of all attempts to form what were termed 'theories of the earth.' A great body of new data was required; the Geological Society of London, founded in 1807, conduced greatly to the attainment of this desirable end. To multiply and record observations and patiently to await results at some future period was the object proposed by them; and it was their favorite maxim that the time was not yet come for a general system of geology but that all must be content for many years to be exclusively engaged in furnishing materials for future generalizations. By acting up to these principles with consistency they in a few years disarmed all prejudice and rescued the science from the imputation of being a dangerous, or at best but a visionary pursuit."

The latter part of the eighteenth century and the first days of the nineteenth were taken up by violent disputes between the two schools of geologists to which Sir Charles Lyell refers. The Neptunists believed all the rocks to have been deposited as layers of mud from a great ocean covering the world. The Vulcanists, on the other hand, saw in the action of volcanoes the prototype of the forces in the past largely responsible for the features of the present world. Such a clash between dogmatic and highly speculative ideas may well be a necessary chapter in the prenatal history of all sciences. But, as both von Zettel and Lyell point out, geology began to progress as a science only when speculative ideas became closely linked to actual observation. At that point, working hypotheses on a grand scale arose which could be linked to predictions of what would be found in the field by a chain of reasoning parallel to those employed in physics and chemistry (see Chapter III).

The work of William Smith of England at the close of the eighteenth century illustrates the way in which a theory and observation can be combined in the study of the earth's surface. This man, an engineer by profession and a geologist by avocation, was the first to classify the strata of England in terms of the mineral constituents of the various layers and the types of fossils which were found. Smith's establishment of a stratigraphic map of England is a landmark in the development of the science of geology. He demonstrated that with the aid of his ideas one could *predict in advance* the relative positions of the different layers in England and in Wales and the nature of the fossils found in each layer. His scheme was both fruitful of new observations and helpful in classifying the observations already made. In short, Smith's stratigraphy meets the criteria we have established for a scientific conceptual scheme.

But to have asked Smith or any other geologist to treat his ideas as purely hypothetical notions without reference to what actually happened in the past would have been to ask the impossible. There was a time when chemists despaired of the reality of atoms; as we have seen (p. 204), they almost abandoned the atomic theory, retaining only a portion of it as a calculating device. There have been great fluctuations in the predominant opinion of geologists as to many conceptual schemes but no skepticism as to the reality of successive events covering a long interval of time. There never seems to have been a geologist of note who failed to believe in the possibility of discovering the order in time of the past events which must have determined the present structure of the earth's crust, and common sense is certainly on the side of this faith of the geologist. Just as any sane man finds it really impossible to doubt the existence of a three-dimensional world and the existence of other people, so too, I think, we would all have to agree that we can hardly doubt that the earth has had a past. This being the case, one may either speculate about this past or try to obtain evidence as to what occurred a very long time ago. Both common sense and scientific methodology lead to the introduction of the time element as a basic ingredient of the conceptual schemes of geologists.

As everyone is well aware, one of the blocks to the early development of geology was the conviction in Christian countries that the Old Testament account of creation must be taken literally. Archbishop Usher in the seventeenth century even calculated the date of the creation and placed it as 4004 B. C. His chronology was still believed by a remarkable number of well-informed and intelligent citizens as late as the first part of the nineteenth century. The account of the deluge was taken quite seriously by many of the first collectors of fossils. That these strange objects represented the remains of former animals was quite generally conceded by the early eighteenth century. But their occurrence was taken often as evidence for the biblical account of the flood. In this area of science, far more than most others, the early speculative ideas were deeply entangled with theological doctrines.

Even today a skeptic might be found to press his doubts about the nature of fossils on a modern paleontologist. He would probably be met with a reply somewhat as follows: that these objects are in fact residual material from previous living plants and animals is proved by the continuity between remains clearly representative of present living plants and animals only recently deposited in the sand or mud, and other fossils of species now unknown. The doubting Thomas would be confronted, for example, with the discovery of the actual flesh and hair of the woolly rhinoceros and the mammoth discovered in the frozen tundra of Siberia. He would hardly be willing to deny what common sense demands, namely, that these specimens were residues of animals which once roamed these plains. From this admission he could be led by gentle stages through a variety of strata of increasing geologic antiquity harboring bones of animals similar to those of the present. That these are the remains of extinct species hardly can be questioned.

The paleontologist, if he were patient enough to have further discourse with the doubter, might go on to connect the location of the strata with the type of fossils found. He could point to many instances where the surface layers contained fossils representing predominantly species still extant, whereas in deeper layers the representatives of known species decreased in a rather regular fashion. Common sense points to the hypothesis that the top layers were deposited last, those underneath at an earlier time. If over the course of time the kinds of plants and animals changed (and the nature of that change is still another story), one would expect the top layers and not the bottom ones to be full of specimens of what we see today. By an accumulation of such evidence, the reasonableness of the basic concepts of stratigraphy of the sedimentary rocks could be demonstrated to any fair-minded group of men. Perhaps even the most cautious analyst of the claims of science would grant that though there were many assumptions involved the observational data obtained in many localities were in general so consistent with these assumptions that they might be considered as proved.

The development of geology, however, has not been all clear sailing any more than that of physics or chemistry. A further inquiry into the history of geological theories in the last 150 years illustrates some of the difficulties. Sir Charles Lyell was a strong proponent of the doctrine known as uniformitarianism which originated with Hutton. He was intent on challenging any idea that in order to explain the present features of the landscape one had to imagine catastrophic events in the past different from any known today. This was part of the reaction against the speculation of the Vulcanists and Neptunists mentioned earlier.

Lyell's emphasis was on wind and water erosion, deposition of layers in lakes and seas, weathering of rocks, and all the phenomena that can be readily demonstrated by careful observation in the field. But Lyell pushed the reaction against catastrophic theories to a point where later geologists have had to retreat from his position. An eminent member of the profession has recently written that "uniformitarianism is not perfectly true for all time," and adds, "We must cling to uniformitarianism so long as our conscience permits." Experimental scientists have similar problems. Pasteur's oversimplification of the relation between life and fermentation comes to mind (p. 229). But when we read in a popular account of geology such statements as "rocks are being made today in the same way as they were made hundreds of millions of years ago," we may easily be misled. As a postulate of uniformitarianism this is a scientific statement comparable to Dalton's assumption that all the atoms of an element are identical. But the layman is all too inclined to treat it as the equivalent of the statement that "red oxide of mercury when heated yields mercury and oxygen" or "George Washington was the first president of the United States." He easily drops into the habit of thinking that this statement of how rocks were formed a hundred million years ago is a scientific fact.

The difficulty in no small measure arises, I believe, because geology appears to be so similar to history. George IV was said to have talked so much about the battle of Waterloo that he came to believe that he had actually been there. An enthusiastic teacher of geology explaining the successive periods of glaciation might well sound as though he had seen it all with his own eyes. Geology expounded as earth history almost inevitably takes on a dogmatic cast. For failure to estimate the degree of probability of the theories of geology and to indicate the extent to which conflicting evidence is as yet unresolved leaves the layman with the feeling that all conceptual schemes stand on the same basis. He appears to be in a position to accept or reject the whole fabric, not as a group of scientific theories but as a factual account of past events. And since the daily papers from time to time announce new hypotheses on some phase of geology or another, an attitude of complete incredulity may well result.

The astonishing popularity of that fantastic book, *Worlds in Collision,* shows how eagerly the reading public welcomes a repudiation of the findings of modern science. The fact that such a volume has found wide distribution in the United States is a distressing phenomenon. Our attempts to give some understanding of science through the formal channels of instruction in school and college obviously have not accomplished all that we might wish. Nonsensical speculation about physics and chemistry today finds relatively little acceptance; any wild notions can be put in their proper place by demanding to know what the consequences are of these alleged revolutionary ideas in terms of new experiments. But the sciences dealing with the past stand before the bar of common sense on a different footing. Therefore, a grotesque account of a period some thousands of years ago is taken seriously though it be built by piling special assumptions on special assumptions, ad hoc hypothesis on ad hoc hypothesis, and tearing apart the fabric of science whenever it appears convenient; the result is a fantasia which is neither history nor science.

Probably a study of geology has more to offer a layman in terms of his own enjoyment than that of any natural science except perhaps portions of systematic biology. This being so, it is more the pity that the dynamic character of this undertaking is not more appreciated. For it is this characteristic that makes geology a science and distinguishes it from history. One could wish that popular articles and books would indicate the distinction between speculative ideas, broad working hypotheses, and well-tested conceptual schemes (though there is, of course, always a hazy boundary between the three types of ideas). The public would then take pleasure in locating the current problems to a greater degree than now and hail with delight instead of viewing with dismay the fact that one theory in geology often rapidly gives place to another. For if this were not the case, the science would not progress, indeed, would be no science at all.

GEOPHYSICS, AN EXPERIMENTAL SCIENCE

Geology is usually regarded as an observational rather than an experimental science. Yet, as in the case of astronomy and biology, its special features do not reside in any distinction between observational and experimental methods. The introduction of long intervals of time into the conceptual schemes is the peculiar characteristic of geology and paleontology. In this century geologists have more and more taken to experimentation and relied on the experimental findings of other sciences. Geophysics is hardly to be distinguished from physics as to the status of hypotheses and theories and experimental procedures. For example, the speed through the earth of waves generated by earthquakes or explosions can be accurately determined. The study of such seismic phenomena is as much a part of physics as observations of the changes of the barometric pressure or humidity, or of variations in the electrical resistance of metallic alloys (though it is worth noting that in the latter case the experimenter has a far wider choice of situations than does the student of the physics of the air; for the meteorologist, like the biologist and geologist, must stick closely to natural phenomena). In all these instances we can use the words "scientific facts" as referring to reproducible experiments; the study of seismic waves seems in every way parallel to a study of the transmission of sound in air or light in a vacuum.

Another example of what might be called earth physics is the interpretation of the small variations in the constant of gravitation which are noted from locality to locality. These can be correlated with the nature of the rock formation and, like the study of seismic phenomena, used to obtain information about the present distribution of solid material; they could be placed in a conceptual scheme from which the time scale had been eliminated. But the same is true of all geological observations. Even the elements of stratigraphy might be formulated without any postulate as to the distant past: the characteristic minerals and fossils of the different layers could be tied together by concepts referring only to observations during the last hundred years at different points on the earth's surface. Though this is historically far from being the way the subject developed and cannot be the view taken by geologists, the possibility of such a transformation throws a light on the methodology of classic geology. William Smith, when he used fossils and rock characteristics for tracing the strata, formulated his concepts in terms of earth history; yet it can be argued that this was unnecessary. He might have introduced a conceptual scheme to assist a classification. In that case he would have been akin not to historians but to the naturalists who in the nineteenth century were so successful in deciphering the life history of plants and animals.

Interestingly enough, in this century physics and chemistry have been introduced into stratigraphy. This fact illustrates a significant aspect of modern geology. Like the experimental biologist, the geologist must see to it that his special concepts and conceptual schemes accommodate the accepted principles of physics and chemistry. Indeed, he may make great use of new developments in these sciences. Thus the discovery of the phenomena of radioactivity in the closing years of the nineteenth century has led within the last fifty years to a whole new subject, sometimes called radiochemistry. This product of experiment and theory made possible by the work of physicists and chemists rests on the conceptual schemes of both sciences. Within the last few decades it has proved highly useful to the geologist. With the aid of certain assumptions which seem reasonable (but are nonetheless assumptions), it is possible to date the various layers of rock by means of special types of analyses of the uranium- or thorium-bearing minerals. What is actually determined is the correlation between stratigraphic observations and chemical analyses. The concepts of chemistry enable one to calculate the age of the strata on the assumption that the rate of the transformations characteristic of radioactivity has been constant over long periods of time and with the further assumption that the mineral analyzed has been undisturbed as to its chemical composition over the same stretch of time as the other constituents of the strata.

The utilization of this method of dating geologic formations has yielded results which show a satisfactory agreement in certain parts of the time scale with former estimates of geologists (made on other assumptions and from other data). Such converging evidence is, of course, of the greatest significance. The total time scale thus revealed has proved to be even vaster than the founders of geology in the nineteenth century had imagined. The oldest rocks containing fossils are considered to have been laid down 500,000,000 years ago and to have undergone extensive changes (metamorphoses) without altering the integrity of the strata. The still older rocks (pre-Cambrian) seem to be at least 1,700,000,000 years old. The age of the earth is now commonly given by astronomers and others concerned with the cosmology of the universe as of the order of 2,000,000,000 years.

One can question whether the principles of physics and chemistry can be applied to the extremely distant past. More than one physicist has expressed grave doubts as to whether over such enormous intervals of time one can assume uniformity as to the behavior of matter. What does the concept of time mean when we speak of thousands of millions of years? Just as the physicist found it necessary to rewrite some ideas about space and time when very high velocities and very small distances came in view, so it is possible that common-sense notions of time cannot be carried over into cosmology. As the data increase which must be fitted into a conceptual scheme with thousands of millions of years as one component, inconsistencies may arise. Such doubts and queries, it will be noted, are quite apart from the question of the validity of the principle of uniformitarianism. This principle is one of those first approximations so often required in the first stages of science; all geological theories agree today in assuming revolutionary, mountain-building periods in which the forces of nature were of a different order of magnitude, at least from those now seen on every hand.

ADVANCES IN THE PRACTICAL ARTS

The success of the application of modern geology, paleontology, and geophysics to practical problems is well illustrated by the rapid improvements in the last thirty years of methods of locating petroleum. Two procedures may be briefly summarized: one depends on the application of the principles of geophysics; the other on paleontology. For example, by measuring characteristics of the speed of seismic waves created by explosions, the geophysicist can often determine the nature of the formation far below the earth's surface. Since the occurrence of certain types of formation has been found to be associated with petroleum, geophysical measurements are of importance in locating oil. The applied paleontologist comes into the story of oil prospecting because of his ability to identify different strata from an examination of the fossil remains of very minute animals. Samples of small bits of the successive layers far below the surface are obtained by suitable borings. If fossils are present, the expert can locate the order of the layers from the specimens observed. Paleogeologic maps thus constructed give a picture of the subsurface structure. Drawing on previous experience as to the correlation of petroleum deposits and stratigraphic data, the oil geologist may be able to predict at what points oil is to be found. The applied paleontologist and the geophysicist in hunting for oil (or minerals) use both empirical observations (and the degree of empiricism is high) and the conceptual schemes of geology. But their success is no more evidence of the reality of what happened millions of years ago than the explosion of an atomic bomb is proof of the reality of neutrons and protons.

CONCERNING THE TACTICS AND STRATEGY OF
GEOLOGISTS AND PALEONTOLOGISTS

However skeptical one may be as to the finality of the conceptual schemes of physics and chemistry and however cautious in the use of the words "fact" and "reality," no physicist or chemist in his laboratory questions the reality of atoms, molecules, electrons, neutrons. (One might add neutrinos, though when pressed even those who use this concept most frequently might admit that instead of being "real" particles, neutrinos may turn out to be a convenient device for "saving" the laws of conservation.) So, too, no geologist, geophysicist, or paleontologist can in his working mood question that his theories are at least an approximate description of what really happened millions of years ago. Therefore, in spite of an apparent fundamental distinction between geology and the other natural sciences, we find that much of what has been said in earlier pages applies equally to the sciences under review in this chapter.

In his excellent book on the history of geology, Sir Archibald Geikie summarized some of the characteristics of the history of that science in the eighteenth and nineteenth centuries. It is interesting to compare his conclusions with what has been revealed by our survey of cases touching on the tactics and strategy of physics, chemistry, and experimental biology. Geikie points out how few of those who forwarded the science of geology up to the end of the nineteenth century were in the strictest sense professional geologists. Almost all the distinguished names either represent men of means who "scorned a life of slothful ease and dedicated themselves and their fortune to the study of the history of the earth," or else were teachers of other branches of science. In short, here as elsewhere in the eighteenth and early nineteenth centuries we find the amateurs playing the leading role in the advance of science. A second conclusion of the same author may be given in his own words: "The history of geological science presents some conspicuous examples of the length of time that may elapse before a fecund idea comes to germinate and bear fruit." This remark emphasizes the recurring phenomenon in the history of all the natural sciences, namely, that the time must be ripe for new ideas to become fruitful or new experience to be appreciated.

Geikie's third and last conclusion is to the effect that "one important lesson to be learned from a review of the successive stages in the foundation and development of geology is the absolute necessity of avoiding dogmatism . . . the Catastrophists had it all their own way until the Uniformitarians got the upper hand, only to be in turn displaced by the Evolutionists. . . . From the very nature of its subject, as I have already remarked," he goes on to say, "geology does not generally admit of the mathematical demonstration of its conclusions. They rest upon a balance of probabilities, but this balance is liable to alteration as facts accumulate or are better understood. Hence what seems to be a well established deduction in one age may be seen to be more or less erroneous in the next. Every year, however, the data on which these inferences are based are more thoroughly comprehended and more rigidly tested. Geology now possesses a large and ever growing body of well ascertained fact which will be destroyed by no discovery of the future but will doubtless be vastly augmented while new light may be cast on many parts of it now supposed to be thoroughly known."

One other similarity between progress in geology and the other sciences we have considered is worthy of special mention: this is the significance of the invention of new instruments and new experimental and observational procedures. We have already noted the prime significance of the comparative studies of the structure of the rocks and the nature of the fossils. Such operations were in essence new observational tools which made possible the development of stratigraphy. The progress of chemistry in the first part of the nineteenth century had made possible a scientific mineralogy (p. 214) without which the study of the mineral constituents of the strata would have had no meaning. Somewhat later the introduction of the petrographic microscope (in the second quarter of the nineteenth century) enabled the examination of the rocks to be put on a still sounder basis. In this century the use of various instruments by the geophysicists has yielded information of importance. Furthermore, the investigation of the behavior of mineral components at high temperatures and pressures in the laboratory has provided a method of checking various assumptions made about what may have happened in the geological past. No further comment is required as to the almost revolutionary consequences of using the phenomena of radioactivity to date the rocks.

The last item in the preceding paragraph illustrates that progress of geology in the last fifty years has been much influenced by advances in other sciences. This is true also, of course, of astronomy, chemistry, and biology; the fabric of the natural sciences now contains so many strands of various sciences woven together that new conceptual schemes must accommodate far more than the data of a special field. (Likewise, new ideas or experimental discoveries may bear fruit in most unexpected quarters.) The consequences of a new working hypothesis on a grand scale in geology may be subject to test in the laboratory nowadays quite as much as by observations in the field; a hypothesis must accommodate in a simple fashion the facts of chemistry and physics and not do violence to the established concepts of those sciences.

Two illustrations of the interrelation of the laboratory and the field may be mentioned by referring to two current problems. One concerns the origin of granite; the other, the origin of petroleum. As to the granite, it is evident that a hypothesis about the formation of a mineral can yield deductions eventuating in limited working hypotheses to be tested by observations in particular localities. The broad hypothesis may also lead to experiments as to the behavior of certain chemical compounds in the laboratory. In either case the facts may be already known; if so, the new idea involves only a reshuffling of observations. But if the hypothesis contributes to the forward advance, new facts will be discovered either in the field or in the laboratory or in both.

In the case of petroleum, the problem leads from geology to organic chemistry and biology. Petroleum is a complex mixture of compounds of carbon and hydrogen. Leaving aside the geologists' dating of the period of the formation of these materials and their present location relative to various strata, we may consider alternative hypotheses as to the source of the carbon. The chemist, accepting the basic postulates of geology, might be able to suggest various origins of this vast and complicated mixture of carbon compounds. Elementary carbon as such might be the precursor, and the intermediates such compounds as calcium carbide, which with water will form the simple hydrocarbon, acetylene. And, given acetylene, the organic chemist has relatively little difficulty in imagining that at high pressures and over long periods of time a diverse mixture of complex hydrocarbons such as petroleum would result. But the use of the imagination in this fashion provides only one more speculative idea.

Another speculation connects the origin of petroleum with the decomposition of animal residue under conditions of high temperature and pressure. A so-called model experiment can be performed in the laboratory in which fish

heated to a temperature well above the boiling point of
water yield a mixture of hydrocarbons superficially somewhat like petroleum. This indicates that there is probably
nothing inconsistent with this particular speculation and
the facts of chemistry. Still another notion about the origin
of petroleum (and one popular today) suggests that algae
of past ages provided the carbon by assimilation of carbon dioxide from the atmosphere by means of sunlight.
A difficulty this idea has to overcome is that today the
result of such synthesis in green plants is protein, fat, and
carbohydrate. One could postulate, of course, that millions
of years ago algae or other plants formed hydrocarbons in
considerable amounts in place of what we consider the
normal products. But this is exactly the type of postulate
that seems fruitless. For since one assumes that there is no
similar process at work today, no laboratory tests are possible, and it is hard to see how such an assumption about a
past biochemical mechanism can be verified by geological
observations.

This problem of the origin of petroleum has been reviewed briefly merely to illustrate the ease with which one
may speculate within the framework of geology and how
difficult it is to form even broad working hypotheses, let
alone develop ideas which can be dignified by the name
of conceptual schemes. It would be out of place to attempt
to summarize the evidence now available as to the origin
of petroleum; but lest I have drawn an unduly confused
picture, I should note that the presence of certain complicated carbon compounds in various samples of petro-
leums indicates (a) that the origin is in whole or in part
from plant or animal tissue, (b) that the material has
probably never been subjected to high temperatures. The
field seems open for ideas which would be fruitful of ex-

periments in the laboratory or new geological observations. Speculations leading to such ideas would be useful; otherwise the free play of imagination can do little more than provide entertainment.

THE ORIGIN AND EVOLUTION OF LIVING ENTITIES

Common sense tells us that there is a beginning to everything, and carrying this idea over into science we generally assume that the universe, this planet, and life must have at some time "originated." Whether this assumption can stand without serious question is at least open to debate. But as one element in a broad working hypothesis it is certainly justified even in the mind of the greatest skeptic. Ideas as to the origin of the universe and this planet fall within the area usually assigned to the astrophysicists and astronomers; to the extent that these scientists are concerned with these and similar problems, they too are working with conceptual schemes in which vast magnitudes of time are an essential component. That theories of cosmology must be consistent with the facts of astronomy and physics is obvious, and as to the need of their being fruitful, there is no necessity for laboring that point again.

Since in this account of the methods of science I have not been able to include a discussion of astronomy, I shall omit any consideration of the difficult problems of cosmology. Rather than question the speculative ideas and hypotheses now current in this field, I shall conclude this chapter with a brief reference to the work of those biologists who are interested in the origin and development of life. Here the contrast between vague speculative ideas and fruitful conceptual schemes can be clearly drawn. About the origin of life we can speculate, but as far as I am aware there have been few ideas put forward that can even be called working hypotheses. Since Darwin's time, on the other hand, evolutionary ideas have become a conceptual scheme, fruitful almost beyond measure.

As to the origin of life, new ideas are a dime a dozen. But how can one transform them into working hypotheses on a grand scale? Consider two widely different notions current today. (They can hardly be designated as more than notions.) On the one hand, there is the older idea that the only carbon compound present on the earth when life began was carbon dioxide (free in the atmosphere or combined in the rocks). Taking this postulate as given, one can speculate as to how it was possible for carbon dioxide to become converted into materials such as sugars and amino acids which are essential for life as we know it. Mechanisms have been suggested that involve those special bacteria which we know can operate as assimilating agents for carbon dioxide without utilizing sunlight. Later, we may imagine, the green coloring matter requisite for photosynthesis was developed; from there on speculation runs into few chemical difficulties. An alternative hypothesis has recently been put forward which as a speculative idea sounds as plausible as the other. It assumes that long before life began there was a vast accumulation on the earth of carbon compounds far more complex than carbon dioxide. If such a mixture included the simple amino acids, aldehydes, and keto acids related to the sugars, it is possible to conceive the formation of complex molecules, such as the green coloring matter of plants, before living organisms developed. But again, what consequences that lead to observation or experiment flow from these ideas? Possibly model experiments in the laboratory; if so, so much the better. But at present, regarding the problem of the origin of life, the cautious reader will register only a series of large question marks.

We know very little, if anything, about this matter, yet I should like to emphasize that it would be highly dogmatic to say that we could never evolve a fruitful conceptual scheme as to the precursors of the most ancient plants. In staking out the limits of our ignorance and insisting on registering the degree of uncertainty of our knowledge, it is essential not to slip into a defeatist attitude about the scientific study of the past. No one, two generations ago, could have predicted that a new kind of chemical analysis of minerals would yield a consistent series of data apparently correlated with the ages of the rocks. No one could have foreseen the interesting use of the radioactivity of one kind of carbon in dating some of the artifacts of primitive man as has recently become possible by the joint labors of chemists and archaeologists.

If scientific studies continue to be pushed forward with as much vigor and freedom in the next fifty years as in the last half-century, the conceptual schemes now used by geologists and cosmologists will almost certainly be greatly altered. For me at least, predicting the fate of a scientific idea for the balance of this century (under the social conditions specified) seems more profitable than speculating as to its truth, if by truth one brings in the notion of a relationship to reality; for a consideration of what is "real" leads to a tangle of philosophical difficulties. From this point of view, I venture to predict that the part of the atomic-molecular theory which has survived the last hundred years will survive the next fifty; as to our ideas of neutrons, protons, and electrons, I should feel less certain. The basis for such a prediction is to be found in the length and nature of the history of the theory in question.

Turning to the study of the past, we may be permitted to guess that the outline of stratigraphy as now accepted will be little altered; but in regard to such questions as the origin of granite, of petroleum, of life, the notions of 1950, I believe, will seem quaint by the year 2000.

To turn from speculative ideas to a fruitful conceptual scheme, one need only jump from the question of the origin of life to the problem of the evolution of the many species of photosynthetic plants and the animals living on such plants. To discuss the impact of Darwin's ideas on the cosmological outlook of the Christian world would be to deviate too far from the purpose of this book. But a few words are in order to indicate the place of the theory of evolution in my analysis of the tactics and strategy of science. The fundamental idea that the species of plants and animals now found had not been fixed once and for all at some distant time was of course not new with Darwin. What Darwin supplied was a series of further working hypotheses on a grand scale as to the mechanism by which changes from one species to another might occur. Therefore, an adequate analysis of the conceptual scheme of evolution as it stood at the close of Darwin's life would have to take into account not one grand working hypothesis or conceptual scheme but several.

Leaving aside the difficulties which the theory of evolution (to give a name to the bundle of concepts) presents to those who take the Christian documents literally, the history of the last hundred years shows that biologists and paleontologists themselves were confronted with many serious problems in developing the new conceptual scheme. These came to a head toward the close of the nineteenth century. As a conceptual scheme Darwin's ideas about what had occurred in the past had to be reconciled with the discoveries of the nineteenth century regarding the mechanism of heredity. The rediscovery toward the close of the century of the fundamental work of Mendel seemed at first to make it extremely difficult to reconcile some of the postulates of the Darwinian scheme of evolution with the findings of genetics. However, in this century, and particularly within the last twenty years, there seems to have been a remarkably successful converging of evidence from paleontology on the one hand and genetics on the other.

To a skeptical inquirer, probably the actual demonstration of how certain bacteria today can accommodate themselves to a changed environment (brought about by such drugs as penicillin) is the most convincing evidence that can be given of the reality of biological change; it makes probable the kind of mechanisms postulated by modern evolutionists to account for the origin of species. Those who are interested in a summary of the modern point of view may be referred to an article written for the layman by one of the leaders in genetical research which appeared in *The Scientific American* for January, 1950. For those who wish to pursue the subject further, I mention again the small book by Julian Huxley, *Heredity, East and West,* and the larger volume by the same author called *Evolution. The Modern Synthesis.* A careful reader of these works will be impressed by the successful convergence of independent lines of evidence, some based on a comparative study of fossil remains, some on breeding experiments with plants and animals; still others are based on the changes brought about in microscopic organisms whose life span is short and which yield many generations in a short period of time.

Therefore, at the moment the evolutionary doctrine seems to stand on a firmer basis than ever before. But it still remains a conceptual scheme which if our analysis of science is right must be judged by (a) its economy in accounting for all the known facts, (b) its fruitfulness in terms of new observations and experiments. *"But is it true?"* the hardy reader may inquire. At the risk of being wearisome, I repeat that the cautious approach to scientific concepts and conceptual schemes would lead one to answer such questions only in terms of predicting probabilities as to the future course of science. My guess would be that fifty years from now what Huxley calls "the modern synthesis," while no longer modern, will be considered a satisfactory step in a profitable direction. In short, the Darwinian revolution will still be regarded, like the Copernican and the Newtonian, as marking the advent of a highly successful new conceptual scheme.

One last paragraph must be inserted to answer a question which has undoubtedly been in the mind of more than one reader. If history and science be totally different kinds of activities, what about archaeology? This subject represents the borderland, would be my reply. At one extreme, it touches science where the skeptic sees only conceptual schemes; at the other, it is a supplement to recorded history where the skeptic questions only the sufficiency of the evidence for the reconstructed picture of the past activities of human beings. Those concerned with what is sometimes called prehistory and protohistory are attempting to handle fragmentary evidence about men who lived some tens of thousands of years ago. The ideas of these scientists do not provide the same kind of information for the lay reader as do the historians' labors. There can be no question of depicting the reactions of human beings to a diversity of problems; the reconstruction of the life of primitive man yields no "accumulation of years to us." Rather the interest is scientific; but as in geology, speculative ideas can gain popular currency far too fast. Broad working hypotheses in prehistoric archaeology have undoubtedly been evolved in the course of the last century, but one may question whether, as far as so-called primitive man is concerned, they have as yet attained the status of conceptual schemes. We may know more about the origin of man than of the origin of life, but I am inclined to doubt it. At all events, both problems from the point of view of common sense are to be regarded by the cautious inquirer as being in the realm of science and not history. But, admittedly, the line separating conceptual schemes in which time is one component from historical knowledge can be drawn wherever one wishes within a large area of archaeology; and the dividing point will depend largely on the faith or skepticism of the individual concerned.

CHAPTER ELEVEN

The Impact of Science on Industry and Medicine

Iɴ ᴛʜᴇ preceding chapter we were concerned with the study of the past, with theory rather than practice, but with a nonetheless highly controversial area. Indeed, the reader's appraisal of what I have there written will depend largely on his or her convictions as to the nature and destiny of man. If I had followed up my slight excursion into history by a chapter devoted to the scientific study of man —anthropology, psychology, sociology—his or her equanimity would have been even more sorely tried. I should have jumped from the frying pan of the past into a roaring present fire. And quite frankly I have no intention of being singed by that particular conflagration. Therefore, only in a few paragraphs of the next and final chapter shall I venture to discuss very tentatively some possible bearing of my analysis of the natural sciences on the scientific study of man as a social animal.

This and the following chapter will be devoted primarily

to a consideration of practical affairs of immediate interest
—the organization and financing of pure and applied research in a free society, and the implications of such arrangements for industry, medicine, and war. The subjects
to be reviewed will turn out to be no less controversial,
however, than those whose examination has just been concluded; the differences in opinion evoked reflect in this
case not theological prejudices but diverse social and political philosophies. My analysis of the present and my
suggestions for the future will be viewed in quite different
lights by a Communist, a left-wing member of the British
Labor party, an orthodox New Dealer (if any be still
extant), a follower of John T. Flynn, or a political agnostic. Indeed, the whole history of science in the last
three hundred years may be read differently by people of
differing political views. For example, starting with a Marxist interpretation of the interaction of science and society,
one group of writers has emphasized a particular set of
historical facts and then proceeded to make radical proposals for the future; both its reading of history and its
suggestions have been challenged by an opposition party.
The battle started in Great Britain just before World
War II; those who are interested in the two conflicting
interpretations of the seventeenth century history are referred to G. N. Clark's *Science and Social Welfare in the
Age of Newton* which reviews the controversy from a non-
Marxist standpoint.

To a considerable degree the debate has turned on (a)
whether there is any justification for a distinction between
pure and applied science and (b) how society should influence the future course of all science. The adherents of
the Marxist philosophy have in general denied the validity
of the idea of science as an undertaking separate and dis-

tinct from the improvement of the practical arts. According to this view, "the advance of the broad front of science . . . is largely governed by its application to current social needs." (The quotation is from a pamphlet, *The Development of Science*, published by the Association of Scientific Workers.) From such an interpretation of the history of science the following conclusion is readily drawn:

"In the postwar world the tasks of reconstruction will set many urgent problems to science *and science will progress only in so far as it plays its part in solving these problems.* That does not imply the abandonment of fundamental research in favor of purely technological work (which would be a reprehensible and suicidal policy for science), but it does imply some degree of planning of the whole field of scientific efforts for the benefit of the community as a whole."

To this statement of the Association of Scientific Workers, F. S. Taylor replies under the banner of the Society for Freedom in Science (*Occasional Pamphlet #1,* April, 1945):

"But what is meant by 'some degree of planning of the whole field of scientific efforts' . . . ? Is some commissar to say, 'These studies of the crystal structure of the phosphotungstates are not related to the benefit of the community—go and do something else.' "

Quite clearly this debate about the proper reading of the history of science and about the place of "pure" research in the modern world, which has been going on in England for more than a decade, is not unrelated to far-reaching questions of a political and economic nature which have likewise been much to the fore in that country. It is also not unrelated to the fundamental problem of freedom for the scientific worker. This in turn brings up the question of how best to organize research in industrial enterprises, in government bureaus, in institutes, and in universities.

THE CHANGING STATUS OF SCIENCE AND INVENTION

My own views on the relation of the practical arts to the origin of modern science have been set forth in an earlier chapter (p. 45). After the birth of the experimental philosophy in the seventeenth century considerable time elapsed before advances in science began appreciably to influence progress in the practical arts. Although seventeenth-century scientists were very hopeful of the practical benefits that would flow from their experiments and their new type of philosophizing, the actual results were far from startling.

Confining our attention for the moment to the physical sciences and their application, it is interesting to follow what occurred in the eighteenth century after the beginnings of the industrial revolution. We note such milestones in the evolution of the iron industry as Smeaton's improvement of blast furnaces using coke for the production of cast iron (1760), the invention of the crucible steel process, Cort's puddling process for making malleable iron (mild steel), the introduction of Watt's steam engine into the foundry (1790's), and conclude by underlining the fact that in 1796 the production of cast iron in Great Britain had reached 125,000 tons, double the figure of a decade earlier. While this industrial revolution was in progress, science was likewise rapidly advancing. The men engaged in both enterprises were in touch with each other, yet the progress of the iron industry and even the development of the steam engine owed but little to the advance of science. Lavoisier's new chemistry was not generally accepted, it will be recalled (p. 190), before the nineties and therefore almost all improvements in iron and steel manufacture were made before the fundamental chemical distinction between cast iron, wrought iron, and steel was recognized (the differences in physical properties reflect differences in the content of carbon). The cut-and-try, empirical methods of practical men were at that time little affected by science; the degree of empiricism in the practical arts was still very nearly a hundred per cent.

The eighteenth century, of course, was dominated by Newton's great synthesis. By combining mechanics and astronomy he had produced a new cosmology that enormously influenced the intellectual world. The entire attitude of educated men toward science had completely changed in the course of a hundred years. Galileo's struggle with the church on the question of the truth or falsity of the heliocentric doctrine was now far distant history. The Royal Society and the French academy had been active for many generations; their scientific journals were recognized media for recording new ideas and new experiments; the meetings of these societies were the occasion for the reading and discussion of many important papers. But one must remember that the number of active investigators in science was infinitesimal by present standards, and they were all essentially amateurs.

By the middle of the nineteenth century, however, the scene changed; as far as pure science is concerned, we seem to have stepped into modern times. Amateurs are still making contributions, but their role is rapidly diminishing. Michael Faraday is carrying on a one-man research institute at the Royal Institution in London (a strange philanthropic foundation of Count Rumford converted into a research laboratory and popular lecture platform by Sir Humphrey Davy). On the continent, professors of chemistry, physics, natural science, medicine, are actively engaged in research. Parliamentary reform is soon to put Oxford and Cambridge in a position to start a new life in which scientific investigation will play a vigorous part. The number of active scientists has increased many fold as compared with a century earlier, and before the third quarter of the century is complete, science will become firmly domesticated in the academic scene in Great Britain, as on the continent, and in the United States.

Most significant of all is the penetration of science into the chemical industries and the founding of electrical industries as a consequence of scientific discoveries. In Germany these industrial developments established by the 1880's a connection which assured an outlet for many highly trained students, holders of the Ph.D.; the professionalization of science was well under way. By the end of the nineteenth century it was the amateur not the professional who was a rare individual in the list of distinguished men of science. Scientific research in universities and to some degree in industry was now a recognized profession; applied science in the forms of civil, mechanical, and electrical engineering was enrolling more and more recruits.

In the eighteenth century the successful inventor had to be both a bold man of business and a skillful empirical experimenter. (Watt was fortunate in finding in Boulton a business partner without whose help he would never have succeeded.) The same is true to a considerable measure as regards the pioneers in the electrical industries that sprang up in the second and third quarters of the nineteenth century. But here the science of electricity provided at least the starting point for experimentation as illustrated by the biography of Siemens (famous for his advances both in electrical engineering and in steelmaking). His life provides an interesting contrast with that of Watt on the one hand or the makers of history of the modern radio industry on the other.

The increasing evidence that the applications of science were going to revolutionize industry and, indeed, everyday life was by no means hailed with joy by those who were advancing science in the second half of the nineteenth century. The previous generation of scientists, men like Liebig, who brought the new chemical discoveries to agriculture, and even Faraday, had been quite willing to move in and out of the applied fields. But an academic aloofness toward industry began to develop in some quarters; the contrast between pure and applied science became the contrast between what was difficult yet noble and what was easy and sordid. In short, many scientists looked down their noses at mere inventors. The inventors often returned the compliment by being derisive toward the impractical and theoretical laboratory men and mathematicians.

There is an amusing illustration of the first type of snobbery in the opening remarks of James Clerk Maxwell's lecture on the telephone at Cambridge University in 1878. Here was a professor famous for his contributions to the advance of pure science, and this is what he said about the revolutionary invention of Alexander Graham Bell:

"When, about two years ago, news came from the other side of the Atlantic that a method had been invented of transmitting, by means of electricity, the articulate sounds of the human voice, so as to be heard hundreds of miles away from the speaker, those of us who had reason to believe that the report had some foundation in fact began to exercise our imaginations in picturing some triumph of constructive skill—something as far surpassing Sir William Thomson's Siphon Recorder in delicacy and intricacy as that is beyond a common bellpull. When at last this little instrument appeared, consisting, as it does, of parts, every one of which is familiar to us, and capable of being put together by an amateur, the disappointment arising from its humble appearance was only partially relieved on finding that it was really able to talk."

Later on in the same lecture Maxwell said, "Now, Professor Graham Bell, the inventor of the telephone, is not an electrician who has found out how to make a tin plate speak, but a speaker, who, to gain his private ends, has become an electrician."

One of the characteristics of the inventor of the classical period of scientific invention, say from 1825 to 1925, was that he could work almost singlehanded. A barn or an attic and a small bank account, coupled with the rare combination of imagination, tough-mindedness, and perseverance, were sufficient. Not so today. By and large, the laboratory of applied research—that is, a group of men highly trained in science and technology—has taken the place of the solitary inventor. This has been the result of a fusion of advance in science and progress in technology that has forced the inventor into partnership with other inventors with different skills and knowledge; the resulting requirements of equipment and supplies have necessitated a new and enlarged method of financing inventions. However you put it, the two social phenomena are as interrelated as good roads and modern cars.

The restrained disdain underlying Maxwell's remarks on Bell is still to be found in a few academic circles; but with the change in the mode of making inventions, social hostility between the pure scientist and the applied scientist (the modern inventor) has almost ceased. The only people who deplore this change—and they do so very quietly—are some who are disturbed lest the process of fusion to which I have referred should mean the elimination of further advances in science itself. That there is a danger in this direction I think every candid student of the recent history of science would readily admit. To offset this danger there must be a persistent and effective campaign to persuade the American public of the importance of supporting research in general.

SCIENCE AND INDUSTRY: THE PRESENT SITUATION

Before attempting to convince the man in the street that more research in pure science is required, we must first analyze the present status of science in industry. In the electrical industry and the chemical industry in the late nineteenth century the research-minded scientist proved his worth. The new pattern was set in Germany and those nations which at that time were within its cultural orbit. Just before and during World War I, organized research as an adjunct to an industrial company began to appear in the United States. Since then the development has been rapid and revolutionary in its consequences. For example, the number of persons employed in industrial research rose from less than 10,000 at the close of World War I to 50,000 at the outbreak of World War II and well over 130,000 in 1949. The total expenditure in this country for research and development work by industry, by the government, in the universities and research institutes has been estimated as something like $160,000,000 in 1930, $350,000,000 in 1940 and half a billion dollars in 1948. The order of magnitude of these figures indicates a transformation within a generation of an enterprise carried on by a handful of men to a social phenomenon of tremendous significance. The lone inventor and the scientific amateur are almost as extinct as the American buffalo.

Statistical information concerning research activities in the United States can be somewhat misleading, for both research and development work are usually lumped together. This is the equivalent of combining under one head science, invention, and the engineering of new industrial developments of the nineteenth century. Today it is convenient to differentiate between (a) fundamental research, (b) applied research, (c) engineering development, (d) production engineering, and (e) service engineering. (I am here following W. R. Maclaurin's terminology as given in his interesting and informative book, *Invention and Innovation in the Radio Industry*.) Under the first heading —fundamental research—we place almost all the activities which we have been considering as science; these may be summarized as the development of new concepts and improvement of older ones (the reduction of the degree of empiricism in a scientific area), as well as exploration with new instruments and new techniques. Applied research has as its goal the application of the existing conceptual schemes to solution of practical problems, the exploration of practical uses of new experimental discoveries, the accumulation of factual information for immediate practical ends. Development work involves the first steps in reducing ideas to industrial practice. The boundary between engineering development and production engineering is hazy; generally speaking, the first concerns pilot plants, the second, improvements in actual large-scale operating units; those engaged in service engineering are in close contact with the sales department and hence the consumer.

Perhaps an analogy from our wartime experiences, in which research played a highly significant role, will serve to clarify some of the relationships—or, better, interrelationships—between the various elements that today combine to effect technological improvements. In describing the vast wartime activities of the army, the navy, and the Office of Scientific Research and Development in designing and producing new instruments of war, one might use the metaphor of a continuous chain running from the laboratory to the battlefield. Along this chain flowed new ideas—ideally without hindrance and in both directions. From the laboratory came many projects to be reworked by the development engineers and, if promising, to be handed on to the manufacturer for production engineering. From the producer came the products which after due testing and inspection were placed in the hands of the ultimate consumer, in this case the fighting man. From him, in turn, came suggestions for modifications of the equipment and ideas for entirely new instruments, weapons, or supplies. Parenthetically, it might be noted that it was no simple matter to keep this reverse flow of information, suggestions, and demands moving smoothly to the development groups and the laboratories.

Not only did information have to pass freely along the chain but decisions had to be made at many points. The requirement of speed at all costs made these decisions peculiarly difficult during the period of hostilities. At the same time, the overriding priority of the needs of the fighting forces made it necessary in many cases to omit those steps which efficiency would have demanded in times of peace. Instead of accumulating information on which one could make a single choice of the best route, several alternatives were often risked at once. A striking example is the production of the atomic bomb. At the outset (the laboratory stage), there were many alternative plans for producing nuclear fuels. As the Smyth report makes evident, several possibilities were still open when the development stage was reached. Conservative procedures would have narrowed the field still further by means of development work before the plans for manufacture were determined. As is well known, instead of waiting for research and development to give adequate information on which to base a rational choice, several lines of approach were simultaneously undertaken with full force.

During the war, time was of the essence; very special conditions were imposed upon all involved, including those connected with science and its application. Therefore, one must be cautious in drawing conclusions from the successes or failures of the wartime period and arguing by analogy. Nevertheless, a consideration of many industrial activities before the war and in the present period will reveal, I believe, a similar chain starting with the research laboratory and ending with the consumer.

Two of the most difficult organizational problems during the war were liaison and communication. Similarly, industrial management is often confronted with serious questions as to how to insure the rapid and effective flow of information between the different links in the continuous chain. Usually the information that the research scientists and the development engineers can provide the executives responsible for policy decisions accumulates slowly. In some cases, therefore, opinions must be formed on the basis of inadequate technical knowledge.

It would be my guess, from having looked over the fence, so to speak, at a variety of research and development problems, that it is a fairly common experience for laymen to be confronted with extremely difficult decisions in just this way. For example, there may be a proposal, highly technical in detail, the acceptance or rejection of which will have far-reaching consequences. The gamble must be assessed both in terms of the probability of success (in essence a technical matter) and in terms of the consequences of success or failure. Hence, the decision involves matters of broad policy; it must be made by men who carry the over-all responsibility, that is, by executives in a position to understand the implications. The proposal may occur at a number of points along the chain; it may be a proposal for a research program, for a development project, for the design and erection of a new plant, for a modification of current practice. Who is to make the decision, and how are the technical factors to be evaluated?

In a smooth-running industrial organization where the whole chain is under the control of one management, those who make these decisions have "grown up with the outfit." Whatever their formal schooling, these men have learned to weigh the evidence presented by the scientific experts and the engineers. A successful executive of this sort almost intuitively avoids the obvious pitfalls into which more than one inexperienced person fell during the war—specifically, by trying to be the expert himself. Likewise, the successful executive unconsciously makes due allowance for the emotional biases of those who claim to know, for he appreciates that scientific opinion is not so cool and dispassionate a thing as the public sometimes thinks; the pride of authorship runs strong and deep, and the wise layman, like a commission of inquiry, does well to balance the personal prejudices of one witness against another. Finally, such businessmen have an understanding of science and its application (at least to their line of work) which makes them realize how important it is to discover the degree of novelty in the new proposal and the degree of empiricism in the science underlying it.

The degree of novelty may reside in the fact that the scientific concepts in a particular field have taken a new form, or it may proceed from new experimental facts in a distant field which are nonetheless relevant to the affair at hand. Or it may be introduced by new materials (alloys and plastics, for example) or by new machines and instruments developed elsewhere. In any event, the executive may well ask the experts, "Why has this never been done before?" Of all the possible answers the one to be most distrusted (though not infallibly the basis for a negative conclusion) is: "No one ever thought of it before." The most convincing, I should imagine, would be: "For the first time we really understand what we are doing"; or "Until last month this experimental fact was not only unknown but totally *unsuspected*."

Roughly speaking, where the degree of empiricism present is high and the undertaking in question has some utilitarian end, we are dealing with the age-old attempt to improve a practical art. On the other hand, where the degree of empiricism is low and the objective is to reduce still further this element in our understanding, we are concerned with a phase of that revolutionary activity which started about 350 years ago and which we designate science. I have suggested that this way of looking at modern research may have a certain value for those who must decide how much to risk on a new enterprise or a new program of research. In general, the more we understand the fundamentals, the more likely we are to succeed in a new scientific or technical endeavor; in short, the lower the degree of empiricism involved, the better.

An example of the practical consequences of the differing degrees of empiricism in two modern scientific fields at a given period may be cited from World War II. The illustration involves two major policy decisions made in Washington during the war in regard to development and production programs. On the one hand, a decision to proceed with a vast expenditure for the production of nuclear fuels for atomic bombs was taken, simply on the basis of laboratory experiments which had been made with submicroscopic quantities of materials. On the other hand, it was decided to continue with the production of penicillin by the biological method of growing molds in cultures rather than to spend money and materials trying to develop a much more efficient synthetic industrial process, even though the organic chemists seemed to be close to agreement on the structural formula and one might suppose an industrial synthesis was just around the corner.

Both decisions were correct. Subsequent events showed that to gamble on the nuclear fuels was right; to gamble on the synthesis of penicillin by the organic chemist would have been fatally wrong (even today no industrial process is yet available). What was the basic difference? The nuclear physicists could make predictions with confidence, and these predictions came true because the interaction of neutrons and nuclei had already been formulated in terms of a conceptual scheme which was adequate; the degree of empiricism in this field, in spite of its recent origin, was low. The element of empiricism in synthetic organic chemistry, on the contrary, was sufficiently high so that few would dare gamble on the success of a commercial synthesis of a complex substance within a short time.

I am quite confident that at no time in any of the discussions concerning nuclear fuels and penicillin was the phrase "degree of empiricism" employed (that is my own invention). But I am inclined to think the basic analysis of the two problems followed the path I have outlined. At all events, the outcome illustrates the point at issue. Proposed advances in the practical arts must be considered in terms of the degree of empiricism of the related science.

PROBLEMS OF ORGANIZATION

It seems quite evident that for efficient operation the chain which runs from the applied research laboratory to the consumer should be under one control. The difficulties of liaison, communication, and decision spoken of earlier are otherwise enormously increased. But the matter is not so simple as merely saying the chain should be under one management. We cannot discuss the organization of research as applied to industry in this way without meeting such fundamental issues as monopoly, governmental regulation, and government ownership.

It is conceivable that for certain reasons, unconnected with technological factors, it may be necessary to have monopoly in a given field (either privately controlled and financed, or governmental). The relation of research to development and to manufacturing is then one of internal organization. With such a monopoly, however, the problem of stimulating the research and the development groups by some kind of technological competition becomes especially acute. And technological competition is as essential for vigor in science and technology as in any other phase of human endeavor.

If, on the other hand, an industry is so organized that there are a multitude of independent producers, then perhaps no one producer can afford a development group or a research laboratory. There are two possible solutions to this difficulty: the government may undertake the development and research function, or a cooperative arrangement involving a large number of competitors may be attempted. One trouble with both these schemes for financing research and development is that the chain is not under one control. Liaison and communication in such cases are extremely difficult, and the executive with authority to make the difficult decisions at various points is usually lacking. Another difficulty is the lack of incentive due to lack of technical and scientific rivalry. Therefore, *if one were considering only technological factors*, the ideal evolution of highly fragmented industries would seem to be a coalescence into half a dozen or so strong rivals. Indeed, I understand this has been the course of history in certain instances and, in part, for the very reason I have named. In any given industrial field a nation would appear to be well off if there is intense technological rivalry between strong competitors, each having control over the entire chain from the applied research laboratory to the consumer.

MEDICINE AND PUBLIC HEALTH: THE SPECTRUM OF THE MEDICAL SCIENCES

This century has been marked not only by a scientific revolution in industry but by a similar one in the ancient art of healing. On the whole, one may say that the penetration of science into medicine is more recent than the corresponding process in industry. The significance in this connection of the work of Louis Pasteur needs no further comment. When this French chemist moved over into the biological sciences he became involved in a branch of science where the practical implications have never been far distant from men's minds (p. 211); his life stands as a great example of one who contributed mightily to the advance of both pure and applied science.

The biological scientist, whether concerned with agriculture or with medicine (and Pasteur was closely associated with both applied fields), has for the last hundred years, at least, moved somewhat more freely from applied science to pure science and vice versa than has the physicist or chemist. This may reflect in part the difference in the relationship of organized society to agriculture, to medicine, and to industry. Governments have been more inclined to assist the agriculturists (including silk-worm cultivators and winegrowers) with technical advice than the industrialists. The monopoly granted to the inventor by government for a period of years through the issuing of a patent finds its parallel, perhaps, in government-sponsored research to provide better knowledge for a whole community of farmers.

Beginning with Davy's treatise on agricultural chemistry of the early nineteenth century, efforts to apply the increasing knowledge of chemistry to agriculture have been persistent in England, on the continent, and in this country. The potato blight and the subsequent Irish famine of the forties focused attention on plant pathology. The new science of microbiology which advanced so rapidly in the hands of Louis Pasteur and other pioneers soon brought many benefits to the agriculturist. In this century genetics has been applied with increasing effectiveness. By the close of the nineteenth century, agricultural experiment stations supported by state and federal money were rapidly making more effective both animal husbandry and the utilization of the soil.

The new discoveries and new concepts in microbiology

were even more important to the surgeon and physician than to those concerned with crops and food products. But it was not until the twentieth century that the labors of the biochemist and the physiologist began to be recognized as forces transforming medicine. Today we can hardly imagine a hospital without a laboratory. The triumphs of the medical sciences have within the course of the last quarter of a century been spectacular; chemists, biochemists, physiologists, bacteriologists have more and more closely cooperated with the clinicians. The degree of empiricism in the treatment of disease has been steadily reduced. New drugs and new procedures, however, have been found by processes still largely empirical. Indeed, in pharmacology, even when rechristened as chemotherapy, the degree of empiricism is high, though in the last decade new concepts have developed and striking experiments have been performed which give promise of being revolutionary.

Parallel to the chain in industry which runs from the applied research laboratory to the consumer is a similar chain connecting the work of the chemist and biologist with the clinician. To vary the metaphor, I suggest we speak of a spectrum of the biological sciences. At one end we place the investigators who are interested only in advancing science; and at the other extreme, the physicians and surgeons concerned with curing patients, as well as the public health men committed to the task of preventing human beings from becoming patients. The equivalents of the applied research laboratories of industry are those located in medical schools and research institutes where biochemists, pharmacologists, physiologists, and bacteriologists carry on their investigations. The engineering development and production engineering groups find their counterpart in those engaged in clinical investigations. And here, too, just as in

industry, it is difficult to draw any hard and fast lines between the different areas; rather, it is of first importance to
see to it that there be close cooperation between those working at these different points on the spectrum.

The analogy I have drawn between industry and medicine may seem inappropriate to the medical scientists,
for they will stoutly maintain that they are as much immersed in pure science as their colleagues in the physical or
chemical laboratories. Even the clinical investigators will
often insist on speaking of their research as fundamental or
basic. Nothing is easier than to start a fruitless debate on
the issue of pure versus applied science, either in the field
of medicine or in many an industrial organization. With the
hope of clarifying the complex issues involved I have employed two metaphors: the chain from a laboratory to consumer for industry, the spectrum of pure and applied research for medicine. In both cases it is fairly easy to identify
those whose chief efforts are spent on an immediate practical object—that is, either making a product or treating a
patient. Likewise, almost everyone will grant that the
theoretical physicist who is evolving mathematical theories of time and space or the interaction of particles has
no other purpose than to advance science. Between these
extremes along the chain or throughout the spectrum, any
single individual may make a case either that he deserves
the support of society because of the practical outcome
of his work or (in another mood) because he is engaged
in basic or fundamental research.

In industry, medicine, and agriculture during the last
two or three generations one and the same individual has

often advanced science quite apart from its applications
and at the same time assisted in the progress of a practical
art through other work. In biology Louis Pasteur provides
the classic example; in physics one may name Lord Kelvin.
When one takes into account the ability of men of genius
to range over the whole field from pure to applied science,
it is not very important to attempt to pin a label on any
given laboratory at a particular moment. The research
laboratory of a modern industry may well partake of the
nature of both a fundamental research laboratory and an
applied research laboratory; so, too, may a research laboratory in a medical school or research institute. The important matter from the point of view of the consumer is the
integrity and vitality of the entire chain; from the point of
view of the patient it is the continuity of the spectrum
which affects his welfare.

At the risk of overloading this chapter with arbitrary
definitions, I should like to introduce one more distinction
in regard to scientific investigation. This I believe to be of
considerable significance for it bears on the support of
science by industry, by philanthropy, and by government.
The amateur scientist of a century ago and the lone inventor were free as the wind in choosing from day to day
the subject on which to focus their intellectual energies.
Even the first research institute (the Royal Institute), since
it was manned by a single individual who was a genius, had
no program. Michael Faraday satisfied the managers by
giving excellent popular lectures and by enhancing the
scientific reputation already established by his predecessor,
Sir Humphrey Davy. Indeed, Faraday's life is an excellent
example of what I propose to call the uncommitted investigator; he ranged from chemistry to physics and all through
the various branches of the latter subject. Pasteur in his

early days illustrates the same independent research man, but as he grew famous and a research institute was equipped for him, he became more and more committed to the application of science to a given end, namely, improving the bodily health of human beings. Social pressures of which he himself was only partially aware closed the gate to his returning to the investigation of those chemical phenomena connected with the optical activity of natural products. An individual or a laboratory which has undertaken to explore a particular segment of a scientific field for whatever reason is committed to a program. The program may be defined broadly or narrowly, by explicit deeds of gift, by contract, or implicitly by the goals proclaimed and the series of victories already won.

Both science and invention in the twentieth century are characterized by an increasing trend toward programmatic research; the uncommitted investigator is becoming a relatively rare social phenomenon. The independent inventor as a potent factor in forwarding the practical arts has in many fields almost disappeared.

Each industrial company, each hospital or research institute, each head of a department in a medical school must decide from year to year on a program, on how widely or narrowly the program is to be circumscribed, on how much fundamental research may be included. C. E. K. Mees differentiates between unipurpose or convergent industrial laboratories and multipurpose or divergent types. But even in the latter there is bound to be some limit to the program. Except as an admitted luxury, few industries would feel justified in advancing science in a far distant field. Likewise, a professor of biochemistry in a medical school would hardly be regarded by his colleagues as "pulling his weight in the boat" if he became interested in some problem in the chemistry of the rare earths and devoted all his energies for years to this particular research. Yet it is quite conceivable that a biochemist might in the course of his work uncover a lead into an area as far removed from biology as my hypothetical example. The important role of the so-called accidental discovery (p. 108) in the past history of the advance of science comes to mind. The amateur scientist was as uncommitted an investigator as one can imagine; and the foundations of modern science were built almost entirely by his hands.

As we review the history of modern science and look at the current scene, it would seem safe to say that the more uncommitted investigators the better, provided they are men of talent and energy. That is to say, the more the better, on the assumption that it is in the interests of the free nations of the world to continue the advance of science. If this assumption be granted, it is worth while examining the forces which tend to increase the emphasis on programmatic research. First, there is the scale and cost of modern experimentation in physics, chemistry, and biology. Few men, singlehanded, with old or borrowed equipment, can make significant contributions. Research teams and large budgets for permanent and expendable equipment are now considered essential by almost all experimentalists. It is not uncommon for an active leader of such a team (the principal investigator we may call him) to have an annual research budget ten times his salary. In terms of manpower alone this means a high concentration of human skills. In terms of dollars the ugly question arises, "Who foots the bill?" Those in control of the funds may be all too inclined to ask: "What is all this money being spent for?" As a consequence even in a laboratory totally divorced from practical objectives, a program may be submitted with a request for funds. Once the expenditure of money becomes justified in terms of specified aims, the investigator is no longer uncommitted: programmatic research is at hand.

The second factor which tends today to increase the amount of programmatic research is the tendency for a research man interested initially only in science to become fascinated with practical problems. This is true in chemistry and physics, and here the locus of the practical application lies in industry. It is particularly true in the medical and biological sciences; there is a constant temptation to move from the pure science end of the spectrum to the applied. There are a number of factors today which operate to emphasize applied research. In some cases they are financial; the investigator's standard of living may be much higher if he devotes himself entirely to applied problems—perhaps because he moves to an industrial laboratory or becomes a consultant with a sizable fee. In other instances, the considerations may likewise be financial but in less personal terms: the size of the research budgets for applied programs are in general greater than those available to men concerned only with advancing science. But more important than motives connected with finances are the subtle forces generated by public opinion. There have been remarkably few scientists over the years who have been completely indifferent to some sort of reward for their labors. (The controversies about priority of discovery alone demonstrate this fact.) Why, after all, should a man in the mid-twentieth century in the United States devote his life to scientific research in areas far removed from any practical application? Unless his work is given false twists by popular-science writers, "blown up" in a way that makes him shudder, who cares whether he succeeds or fails in his endeavors?

The layman often fails to remember that men who devote their lives to advancing science without any thought of applications are engaged in a tremendous gamble. Their success or failure turns on the formulation of new concepts or widening conceptual schemes or making some experimental discovery that will be fruitful not in practical but in conceptual terms. Of this great enterprise of reducing the degree of empiricism in our knowledge (as distinct from our practice), it may well be said that "many are called but few are chosen." Only those who have spent many years in this type of scientific work can fully understand how great is the risk and what are the emotional consequences of that risk. No wonder that in the last decade we have seen a continual migration of talented research workers from one end of the spectrum to the other; the chances of a lucky strike seem better in the applied fields, and the immediate rewards in terms of recognition are certainly far better.

We must recognize, of course, that in certain circumstances the migration just referred to is of importance for society. We need first-class men all along the industrial chain and distributed over the whole spectrum of the medical sciences; and the training of the research worker is such that he is almost certain to start with an interest in purely scientific problems. But if it be true, as I believe history shows, that the significant revolutions, the germinal ideas, have come from the uncommitted investigator, then the present trend holds grave dangers for the future of science in the United States. It is all very well to say that an able man even if committed to a program of research will quickly turn aside to follow a promising new clue into other fields. But the record seems to indicate the contrary. The failure of a conscientious and able chemist, the head of a government bureau, to pursue a trail that led to the discovery of the rare gases will be recalled. His own statement (p. 120) as to the reasons why he was not the discoverer has significance. "Chance favors the prepared mind," but only the mind prepared by a complex process of social conditioning is ready to gamble heavily on following up a clue which at best can lead only to theoretical knowledge.

THE ROLE OF THE UNIVERSITIES

Of course, one must freely grant that there are many individuals who make good members of a research team who have not the special endowments necessary for the scientific pioneer. Undoubtedly over the years many uncommitted investigators who have made little or no contribution to science but only increased the amount of available information would have been more productive if their efforts had been part of a well-conceived program. But if science were too individualistic and anarchic fifty years ago, the danger now lies in the opposite direction. To some degree, industrial laboratories may sometimes afford shelter to a man who will turn the "unexpected corner" in terms of theoretical science. The same is true of research institutes which are committed to broad programs. But by and large the home of the uncommitted investigator must be the universities. As I am obviously a highly prejudiced witness in this matter, I shall cite the testimony of a successful director of industrial research, Dr. C. E. K. Mees, vice-president in charge of research of the Eastman Kodak Company, in his book *The Organization of Industrial Scientific Research* (1950). Referring to the production of scientific knowledge, he writes as follows:

"The basic institution on which everything else depends is the scientific department of the university, and this differs from all other institutions in that it has and should have no direction from outside, and complete freedom in its choice of subject. It is from the universities that most of the new ideas by which science is advanced are likely to come, since in all other institutions there is some restriction and will probably always be some restriction in the fields selected for work."

One may supplement this concise statement of the role of the universities by pointing out that by long tradition a professor in a university is a free agent insofar as his scholarly activities are concerned. As soon as he obtains a permanent position, he is a life member of a community of scholars whose aim is teaching and the advancement of knowledge. He is under definite obligations to perform his share of the teaching service and under a moral obligation to advance knowledge *as he sees best*. This means that if he runs into a blind alley in his thinking and gets stuck there for a year, a decade, or a lifetime it is his own affair. His partners, the other permanent members of the faculty, will regret his sterility, but nothing can or should be done about it. And since his obligation, actual and moral, is dual, he may take comfort in the success of his teaching. More than one unsuccessful (or unlucky) investigator has compensated in this way for his deficiency. That this is possible is one of the reasons why universities are still the only centers where a man interested exclusively in pure science can be fairly certain of finding in one way or another a satisfactory life.

These considerations apply to all permanent members of a university staff. Yet pressures may well develop constraining to some degree the research activities of these investigators; indeed, they begin to manifest themselves as soon as expensive equipment and many assistants are required. Whether this money comes from the university itself or whether it is in the form of outside grants from a private source or the government, it is often tied up with some "project" the professor has put forward. He is under a definite commitment to continue along the line outlined in the project and under pressure to get results. Such commitments and pressures are not necessarily bad, but they do limit the scientist's freedom. They may or may not tend to make him move toward applied research, but they often do. A number of cases could be cited to demonstrate this fact. It is worth noting that the astronomers years ago seem to have found a satisfactory way of carrying on group research and using very expensive instruments without unduly constraining the investigator. The pressure to undertake applied research in their case, however, is missing.

From the foregoing it follows that university careers should be made as attractive as possible. For the keen scientist this means providing excellent opportunities for doing the kind of scientific work he wants at any time. He should have the maximum of flexibility in his plan, the minimum of obligation to carry out a program. Grants-in-aid should be made not for a project but for a man. To those who administer funds to support research, let it be continually said: "Don't appraise the project but the proposed investigator; don't bet on the subject but on the man. And like all successful gamblers, play your hunches heavily; don't distribute your money in small amounts over a wide field."

To my mind, the pendulum is in danger of swinging too far toward organized programmatic research. To any who are worried lest money flowing without control will lead to scientific "boondoggling," one may point out that in science as in industry healthy competition will supply the corrective force. Both in the pure and the applied aspects of research it is of prime importance to have many strong centers (medical schools, hospitals, university laboratories, research institutes, and experiment stations). Under good management each of these centers will endeavor to find the most promising investigators for positions of responsibility and leadership. If this spirit prevails, money flowing from an outside source to these centers is in effect the placing of bets on men and not on programs. More and more the philanthropic foundations have been adopting this policy, at least in the fundamental sciences of chemistry, physics, and biology. In so doing they have been following consciously or unconsciously the pattern established in Germany in the great period of its science (1850–1933). The intense intellectual competition between a dozen or more universities was a potent factor in bringing the German nation into its lead in science. That there were other unfortunate aspects of advanced education in imperial Germany I am well aware. But from the point of view of advancing knowledge, no climate of opinion can compare with what was to be found among the German-speaking peoples of the nineteenth century.

WHY MORE SCIENCE?

By and large the American public seems convinced as to the necessity of spending large sums of money on research to benefit industry, medicine, or the preparation for war. (As for the last, more will be said in the next chapter.) But among those who control the decisions on the flow of both private philanthropy and government funds, there is frequently either a reluctance to support fundamental research or a confusion as to the difference between pure and applied science. The reasons for this confusion have already been referred to; they have resulted from the fusion of science and invention in modern industry.

There are a number of quite separate lines of argument in favor of the support of scientific inquiries by a free society. The scientist himself is inclined to argue that a civilized nation will wish to support its men of learning for the same reasons that the Renaissance princes supported artists and writers. Such a line of reasoning carries on the spirit of the amateur scientist of the seventeenth and eighteenth centuries and comes at times very close to maintaining the doctrine of "art for art's sake." For certain types of individuals, adherence to such beliefs is probably essential for their own successful functioning as investigators. And I should not wish to take the negative in a debate on this point. But while society may be willing to tolerate and even applaud those who pursue science for its own sake, supplying funds on the scale now required for some types of pure research is another story.

A study of the history of the last hundred years provides arguments most appealing to the hard-boiled citizen who looks at all expenditures with a cautious eye. The record is quite clear: from the labors of those who were interested only in advancing science have come the ideas, the discoveries, the new instruments, which have created new industries and transformed old ones. To continue the phraseology of this book, "lowering the degree of empiricism" even in pure science has eventually paid dividends in technology. As to the new discoveries, one need only be reminded that the electrical industry is based on the phenomena of electromagnetism, first revealed by the experiments of the scientists in the early nineteenth century. But again the public can be confused between new discoveries made in connection with advancing science and new inventions, because today the applied research laboratory and the engineering development division of an industry largely take the place of the inventor. Money for applied research? Why of course, your critic may say, but why more funds for persons who are not in the least interested in the application of science to industry, to medicine, to agriculture, or to national defense?

The answer again is to be found in the recent history of many an industry. The applied scientist utilizes time and time again the new findings of those investigators working solely to advance science; furthermore, after a time his efforts to reduce the degree of empiricism in his strictly limited applied area run into a dead end. New concepts and conceptual schemes or new instruments and procedures are required; nine times out of ten these must come from the laboratory devoted to fundamental science. The engineer must call on the applied scientist, and sooner or later, if advance in pure science ceases, he will call in vain. For the applied scientist will run out of his most precious fuel— new ideas and new experimental results.

Maclaurin's account of the radio industry illustrates the interplay of science and technology in one field which is characteristic of this century. The opening chapters of F. A. Howard's account of the development of the oil and synthetic rubber industry (*Buna Rubber: The Birth of an Industry*) tell the same story with chemistry instead of physics as the fundamental science involved. Surely it is no accident that Germany both produced a predominant number of Nobel Prize winners in chemistry and pioneered the production of synthetic oil from coal and the manufacture of synthetic rubber. Pure and applied organic chemistry went hand in hand in Germany from 1860 to

World War II. If the past is any indication of the future, a nation in order to lead in technology must lead in pure science. That in a few words is one compelling answer to the question, why more science?

CHAPTER TWELVE

Science, Invention, and the State

T HE IMPACT of science on industry and medicine in this century has had far-reaching consequences in the field of politics. What was once almost entirely an affair for private organizations has become more and more a concern of the state. Public opinion in the democracies has become increasingly interested in science and invention, and in totalitarian lands some rulers have been keenly aware of the significance of scientific investigations. The vast sums spent for research by the government agencies in the United States during World War II set a pattern which bids fair to revolutionize the American scene. The justification during the war was, of course, forwarding the military program (which included far more than weapon production, for medical research was of great significance for the armed forces). In the uneasy days since 1945, taxpayers' money has been used to support research and development on a scale that is enormous as compared to the prewar period. Most of it has been spent on engineering development or production engineering. Many millions of dollars annually, however, have supported programs of applied and fundamental research in universities and independent research institutes. Among the government agencies controlling these funds, three have played a leading part: the National Defense Establishment (army, navy, air force), the Atomic Energy Commission, and the Public Health Service. To those must now be added the National Science Foundation established by act of Congress in the spring of 1950.

In the present age, government officials can hardly escape having a deep concern for science and its applications. Public health studies, medical research, agricultural experiments will be encouraged in one way or another. Expenditures of taxpayers' money on applied research and engineering development in the industrial field are far more controversial. A nation determined to nationalize many of its industries would take one attitude, a country committed to private ownership another. If a free society decides to place under public control a number of its important industries, as appears to be the case in Great Britain, clearly the government by virtue of this decision becomes involved in the management of industrial research and development. A friendly observer from this side of the Atlantic may wonder how under conditions of nationalization the equivalent of technological competition will be provided. What incentives and rewards can be contrived to stimulate invention and innovation in a highly socialized economy? These are interesting questions. They lead on to other questions connected with the control of applied research by governmental agencies and, eventually, to the relating of the state to industry. To the Marxist the future relation of science and society is envisioned in one set of terms; to those who think in terms of a profit-and-loss economy, in quite another.

Under the competitive conditions which have prevailed by and large in the western world since the beginning of the industrial revolution, the role of government has been confined to affording patent protection for a period of years for certain new enterprises. The significance of patents and the recurring difficulties of the patent system would provide continuing themes for a series of technological case histories. One might start with Watt and his patents for his engine, continue through various nineteenth-century inventions, and end with the two twentieth-century cases already mentioned—the radio industry and the synthetic rubber industry. The story would show, I feel sure, that with all the difficulties and abuses inherent in the patent system, patents have been an essential element in the development of modern industry. That the system could be improved few have questioned, but to obtain agreement on any changes has proved extremely difficult. And any informed public discussion of the issue is almost precluded by the legal intricacies and the mass of technical information involved in any example cited.

Basically, of course, a patent is a highly restricted monopoly granted to an inventor for a period of years. Without the protection afforded by such a monopoly many an invention would have remained a paper scheme: the capital requisite for turning the invention into a reality would hardly have been forthcoming without governmental protection for both the inventor and the innovator (to use Maclaurin's phrase for the businessman who gambles on a new invention). Invention, to be sure, has now largely ceased to be a one-man affair and is usually the consequence of a team effort of scientists and engineers; therefore, the protection is given to a company rather than to the proverbial pioneer in the attic. The advantages of such protection (as well as some of the complications) are well illustrated by the growth of the radio industry and the development of synthetic rubber. I refer the reader interested in exploring further the intricate problems of patents and patent protection in the mid-twentieth century to a study of these two examples as set forth in the books by Maclaurin and Howard already mentioned.

A description of the invention is made public, of course, when the patent is granted. Consequently there has grown up an enormous patent literature. Much of the information thus available to the scientific and technical world, however, is of little value. The patent may deal with trivial or obsolete matters, the description may be inadequate and at times purposely misleading. Few of the checks and balances which have made the scientific literature increasingly reliable operate as to patents; established standards of reporting, careful editing, concern for a scientific reputation—all these are largely absent. In chemistry, at least, no one would care to assert a "fact" based solely on a patent claim. Nevertheless, the patent literature is of value—of greater value in some fields than others. No one working in an applied industrial field would fail to follow the patents which are published. A fabric of public technology, so to speak, has been built up by means of patents published all over the world (though reports from companies *after* a process is in operation often tell far more than the bewildering details of the patents which surround the process). All the technical ramifications, all the "know-how" are rarely proclaimed by any company, however. There are strong traditions of secrecy in many industrial fields. But if there were no patent system, a company would be forced into as complete secrecy as possible to protect the inventions of its research and development departments. That the public would thereby suffer seems quite clear. There can be no doubt that secrecy is basically incompatible with scientific progress, and technical advance is today closely inter-meshed with science.

The patent system has been historically the method used by organized society to encourage industrial innovations. What other responsibilities has a modern state for fostering applied science? Such long-range projects as the utilization of solar energy or the underground gasification of coal or the use of atomic energy for industrial purposes might well be subsidized even by a nation determined to set its face against any trend toward the public ownership of the "tools of production." Yet even in these cases, the degree of subsidy, the nature of the control over the re-search and development programs provide subjects for end-less debate. Nor can such debate be carried on without reference to far-reaching economic, social, and political considerations. In the last analysis, expenditures for re-search and development can be judged only with reference to the ultimate objectives; this is as true of a nation as of a business concern. If we were living in an era of disarma-ment and peace, therefore, government support of indus-trial research might be a highly controversial matter in the United States. In these grim years, such questions sink into insignificance. The overriding priority is now effective and rapid rearmament. And as long as the world remains di-vided into two armed camps, one must judge government policy primarily in terms of the international scene.

Perhaps before these words find their way into print we shall be engaged in a third global war; if so, much of what follows will be obsolete. But if on the other hand, as I believe far more likely, the Soviet Union and the United States remain nominally at peace, we shall for some years be faced with the necessity of keeping the free world heavily armed. Even when we have reached the degree of rearmament deemed satisfactory for the defense of western Europe, the expenditures on national defense cannot cease. We must continue to spend money on a vast scale and much of it for new equipment. In this century major technological changes affecting war come rapidly, and weapons quickly become obsolete. Therefore, in discussing the role of government in science and invention, one must bow before the tragic tensions of the times, reverse the traditional peaceful attitude of the American people, and start the analysis by considering military requirements.

SCIENCE AND NATIONAL DEFENSE

Let us begin by referring for a moment to World War II. At the height of the struggle one could clearly envisage the continuous chain running from the laboratory to the battlefield. Many of the links were analogous to those we have met in describing industrial research—applied research, engineering development, production engineering, service engineering. For a sales department and a peaceful consumer, however, we must substitute in this case the army, navy, and air force engaged in battle. As I write, the chain runs to a somewhat special battlefield (Korea) and to several quite different potential battlefields elsewhere, in the air, on the land, and under the sea. To the existence of potential, not actual, battlefields can be traced many of the problems which beset those in Washington today who must make decisions.

An industry is continuously selling its products; it receives constant information from the consumers, and in the light of this information, fed back along the chain, revises its production, its engineering, and even some of the goals of its research. When the battlefields are actual, the government likewise receives reports from the front on which to base its plans, but when many of the combat areas exist only on paper, those in control must imagine what will be the performance of their weapons. Field tests and various proving devices may tell something, but I doubt if any military man will challenge the statement that only "the battle is the payoff."

In modern war there is no typical battlefield to be imagined, nor is the equipment of the enemy a static affair. The experience in Korea has brought these points home to the American public with no little force. If we must put our troops in action elsewhere in the coming years, under what conditions will they fight and what will be the technological status of the enemy? If the global catastrophe does occur and we fight a world-wide war, where will be the battlefields, what will be the effectiveness of the Soviet Union's improved weapons? These technological unknowns appear to complicate beyond measure the problems of those who must make war plans even in days of relative peace. Weapons in existence are one thing, those on a production line another, those in engineering development still a third, those on a drawing board still more distant; and beyond this segment of the chain lie perhaps revolutionary new discoveries in the laboratory. All these factors exist today to bedevil military planners. Nor must we forget that behind a curtain of secrecy a potential enemy has his own chain of science and technology. No one can be certain what will be the performance of one set of "drawing-board weapons" against another.

I need press the point no further. Vannevar Bush in his

well-known book, *Modern Arms and Free Men* (1949),
surveyed the future of technological development affecting
weapons and military tasks. What I am concerned with
here is solely the matters of organization and of management. Politics in the larger sense of the word, rather than
technology, is involved. The basic problems are as old as
attempts by human beings to organize society; many of
the specific questions have been asked by political philosophers observing this country since the founding of the re-
public. The compromise between the system of checks and
balances established by men fearful of centralized power,
the traditions of a parliamentary system gradually im-
ported from Great Britain, and the demands of a modern
state have produced a form of federal government that
defies description. To some visitors in Washington there are
moments when it seems a madhouse. Yet it works, particularly in times of stress, in a way to defy all the predictions of
an analyst of governmental forms. Since the close of World
War II, however, new types of problems have arisen for
which there has been evolved as yet no adequate political
machinery. Customs and traditions are lacking to assist
those in power to operate effectively the United States
government as a controlling force over the chain of research and development which leads to new weapons and
equipment for fighting men.

THE PROBLEM OF EVALUATING RESEARCH
DIRECTED TO WEAPON PRODUCTION

The management of a progressive industrial company
with a large investment in research and development must
constantly make vital decisions about scientific and technical matters; those in positions of responsibility must establish priorities, dare to decide to drop one line of work, push

another, scrap one pilot plant, build another. Over the years the strong companies have developed groups of men capable of making shrewd guesses as to the future and robust decisions. No comparable human organizations exist today in the federal government; what is more important, there is no tradition of how technical information should be assessed. Questions of the highest import to the future of our military strength have been settled under the duress of social pressures almost unknown to industrial executives. Political forces (and I do not mean party politics) are as certain in the government of a democracy as the gravitational attraction of the earth. Therefore, the mode of operating an industry cannot be applied without profound modification to governmental agencies. Congress controls the funds; the executive, within the limits of congressional consent, controls the immediate decisions through a long chain of command. Public opinion, often ill informed by "leaks" to the press (calculated or accidental), may become so inflamed as to rule out certain alternatives that otherwise seem attractive. All this being a necessary consequence of our free society it would seem important, if we are to face a long period of vast expenditure for weapons and military supplies, to improve the modes of evaluating the results of research and development. We need to develop sound traditions of control over the technological aspects of long-range military planning.

Nothing in the foregoing is intended to reflect on the past or present management of either the Atomic Energy Commission or the Research and Development Board of the Defense Establishment. Within the framework inherited from the past policy of federal agencies and from the war, the men who have manned these agencies have been more effective than one could dare hope was possible.

But few of those who have followed the development of military equipment would deny that we could improve the methods of evaluating the vast amount of technical information today essential to military planning. Above all there is a need for establishing firm technical decisions rather than attempting too much with too little. Political forces almost automatically generate compromise decisions at every level. Those who must make policy cannot bring the wisest judgment to bear on scientific or engineering matters unless protected from pressures generated by outside experts or interested parties.

There is no need for any drastic reorganization of the handling of research and development within the government. To be sure, better arrangements might be possible in some areas, but redrawing of organizational charts and reshuffling of lines of authority slow down any effort, and this is no time for delay. What I should argue for is a revised attitude on the part of politicians, officials, and military men as to the relation of research and development to the production of weapons. The men who now have responsibility in a variety of posts, high and low, should have more real authority and be far less subject to extraneous forces.

What is needed, it seems to me, is the rapid development of a tradition of quasi-judicial review throughout this whole area. When a question comes up to be settled, even if three or four echelons below the top, one or more referees might hear the arguments pro and con. If there are no contrary arguments, some technical expert should be appointed to speak on behalf of the taxpayer *against* the proposed research or development. Then adequate briefs of the two or more sides should be prepared (*not* a compromise committee report). With opposing briefs, arguments, and cross-questioning, many facets of the problem, many prejudices of the witnesses will be brought out into the open. The referees could then report their findings to those who have the responsibility for decisions and the latter would be in a position to give unequivocal answers based on adequate documentation. Such decisions when they reached a higher echelon would be unlikely to be reversed without adequate reason; the briefs and counterbriefs would be there to show that all the relevant arguments had been explored.

Science perhaps may be said to deal with accurate predictions, but not applied science. Fallible human judgments are involved. Technical decisions represent the weighing of all probabilities and the discounting of prejudices. Therefore, if a quasi-judicial procedure were established, few weak compromise decisions would result. Too often now, the conflict of prognostications of experts is resolved by "splitting the difference." Consequently there is a failure to provide adequate support for either of two opposing views.

In what has just been written I must plead guilty to riding a hobby—the need for the introduction of the traditions of quasi-judicial review into government technical programs. But quite apart from the merits or demerits of this suggestion, the questions raised may serve to emphasize the extent to which every citizen in the United States is a party to an enormous new enterprise. His government has gone into the research and development business on a scale totally different from anything seen in the past (except during the actual conduct of World War II). Consequences of tremendous significance in terms of survival may hang on the way this work is carried on. The waste of enormous sums of money could threaten the soundness of our economy; failure to support adequately some areas, on the other hand, might result in our falling far behind in an armament race.

The whole chain from pure science to the battlefield (actual or potential) is a responsibility of the representatives of the American electorate. Intelligent criticism and understanding public opinion are essential if all links in the chain are to operate effectively, particularly in periods of stress.

FEDERAL FUNDS FOR FUNDAMENTAL RESEARCH

Let me now consider the other end of the chain—fundamental research. In so doing one must first recall the impressive record of the role played in modern industry by pure science and realize that in terms of a long-range national defense program, the advance of science is of basic importance. Such being the case, the official representatives of the American people must keep a watchful eye on fundamental research and devise ways and means of encouraging the advance of science, taking care, however, that this solicitude does not have a reverse effect and end by hampering rather than fostering the creative efforts of original minds. If the uncommitted investigator is the key person I believe him to be, then he must be the focus of attention. To the extent that public money is to be spent on encouraging fundamental research, it would seem important, as I have indicated, to support men, not projects, and one may hope this will be the policy of the newly created National Science Foundation. But such a policy while easy to state is difficult to administer, particularly in a period of tension and rearmament. For almost all the political and social pressures act in a contrary direction. Money will be spent freely in the next decade on applied research and development in government laboratories, in industry, and through contracts with universities. Of this we may be sure. But whether fundamental research will thrive is open to serious question. Yet one can hardly overemphasize the importance of continuing pure science even in the years of a heavily armed truce.

That the taxpayer should assist in the advance of science I have no doubt. But it would be disastrous if the sole support of science were from federal funds. Any decision by those in control of other sources of money for research to "pull out of the natural science field and leave it to Uncle Sam" would seem most unwise. Experience over the last decade has shown that private philanthropy can play an enormously important role even in areas where public money is flowing freely; cancer research is a good example. Independent agencies having money at their disposal provide a valuable supplementary source of funds even for investigators supported largely by money from Washington. As a consequence they exercise great influence. Governmental policy is affected by their outlook and the independence of their judgment. The officers and trustees of private funds can even at times give a lead to government officials harassed by petty political forces and by so doing assist in the wise expenditure of federal funds. The future of the uncommitted scientific investigator is best assured by providing him with a bow of several strings; and only one of these should be money voted by the Congress of the United States.

SCIENCE AND POLITICS

In the preceding discussion of the role of government I have mentioned only two aspects of society's interest in the progress of science and technology, the production of instruments of war and the promotion of research in the natural sciences in one country, the United States. The omissions reflect both the temper of the times and the author's economic and political prejudices. In a more peaceful

era some reference would be in order to the research programs directly supported by the federal government in its
own laboratories for nonmilitary purposes. The vast agricultural research activities of the last fifty years might well
be passed in review; likewise the important work of the
Bureau of Standards, the Geological Survey, and the recently expanded activities of the Public Health Service. To
the extent that the scientific activities of all these branches
of the federal government are of direct benefit to the welfare of the nation as a whole and to the extent that they
cannot be effectively carried out by state or independent
agencies, they deserve the support of all citizens. But I
must confess to grave doubts of the wisdom of the expansion of governmental laboratories on the scale of the last
ten years. I question whether the record supports any contention that a government laboratory is a favorable spot
for fundamental investigations; furthermore, applied research looking toward industrial development can be best
fostered by industry itself.

Another obvious omission has been my failure hitherto
to discuss the social sciences. Government support of the
physical and biological sciences has been urged in terms of
national defense and the impact of science on industry and
medicine. How about psychology, sociology, anthropology?
Is it not just as important to have these fields of study
flourish? Perhaps more important, some may say, for the
advances in technology have had a disruptive effect on
the body politic, and more talent might well be devoted
to a study of social and political problems. Few would
deny that the society that has been developed in this
country is unique. It is similar to other democracies in
many respects, but we have certain ideals that are a product of our past. Our solidarity as a nation depends on our

acceptance of these ideals and on a concerted effort to move continuously toward the social goals implied. This is no easy matter, for the complexities of modern society are great. The question, therefore, at once arises whether the studies of man and society by competent scholars can provide basic information of practical value. Can the degree of empiricism in politics (using that word in its broadest sense) be reduced by scientific studies? If so, will the resulting introduction of science into the practical art of organizing human society be of benefit to this free nation?

My answer to these questions is in the affirmative. My confidence as to what may be accomplished flows from the assumption that advances in the science of man and society will go hand in hand with fruitful applications of the present techniques and concepts. Perhaps few people realize how much progress has been made in the last decade and what techniques are now available to help in the solution of practical, human problems. But it is the future which holds promise, for I imagine that even the most enthusiastic psychologist or anthropologist would readily admit that at the present moment the conceptual schemes at his disposal are the equivalent of what chemists and physicists were using in the late eighteenth century. As a consequence, in every case of the practical application of the sciences concerned with human behavior, the degree of empiricism is very high.

As in the medical sciences in recent years, many of those immersed in practical problems should be the very ones to advance the science. If a chemist may be bold enough to give advice to colleagues in a far-distant field, I would say that the life of Pasteur affords the relevant analogy. As an applied scientist, he solved immediate problems and at the same time reduced the degree of empiricism in those branches of biology he made his own. It is easy to be misled by the divorce of theory and practice which characterized the work of Newton, Clerk Maxwell, and even Darwin. There are times and places when even the purest science cannot make progress in an ivory tower. On the other hand, one may express the hope that the ultimate consumer, the general public, and the practical men will not press too strongly for immediate results or fail to recognize that no one can be certain of how much progress can be made even in half a century. Projects concerned with applying the knowledge now at hand must not be so numerous or exacting as to prevent a sturdy effort to decrease the degree of empiricism in the methods now in use. The long-range programs require adequate support and patient waiting. This support might well be provided in part by government funds. For our free society has more need, perhaps, for an increased understanding of the fundamentals of human nature than any other. The empiricism of the past may be sufficient for the masters of a police state, but a free people in this modern age requires as much assistance as possible from advances in the social sciences.

We should have no illusion that basic national issues can be handled by any group of social scientists in the way that problems of design of bridges and machines can be treated by engineers. Policy questions must be resolved, in the future as in the past, by governmental officials, business executives, and labor leaders; they cannot be handed over to scientific experts to find an answer which is "right." Most of the decisions must be made on the basis of experience coupled with the advice of competent analysts and those whom I venture to call social philosophers. History as an extension of experience (p. 266) will continue to provide the chief guide for the administrator and policy maker. Nonetheless the point of view of the social scientists may be of some immediate assistance. The types of problems where one can hope for help from the social psychologist, sociologist, and anthropologist involve human relations and those conflicts among individuals and groups which have been so much intensified by the conditions of modern life. The people of the United States will be the beneficiaries of whatever advances can be made in the study of man as a social animal.

VALUE JUDGMENTS AND
THE SOCIAL SCIENTIST

If the social scientist may be regarded as one concerned with reducing the degree of empiricism in politics (still using the word in its broadest sense), he is analogous to the medical scientist in many respects. But how about human ambitions, social goals, and ethical considerations, some one may ask. Science is neutral as to value judgments, it may be claimed, and such judgments lie close to the heart of political and social problems.

But let us examine the statement which is often made that science is neutral as far as judgments of value are concerned. Is not this one of those three-quarter truths fully as dangerous as half truths? Let us consider the medical sciences today. Investigators and practitioners concerned with human disease almost unconsciously accept a set of values which limits their activity on the one hand and on the other serves as a powerful spur to their endeavors. This fact seems to be overlooked when the neutrality of science is proclaimed. Much more than the Hippocratic oath is involved. Only in a society where life is considered preferable to death and where health is glorified would funds flow freely for the study of disease. Only where the sanctity of each individual is so strongly felt that it is regarded as a paramount duty to save every life possible at whatever cost would physicians, surgeons, and medical scientists act as ours do today. Our standard of medical care and our desire to raise it are based on a series of value judgments. Let me make it plain—I am not questioning the assumptions. I am merely pointing to the existence of these postulates basic to all work in the medical sciences. I do so because I believe the situation is analogous in the case of those scientists who are investigating human behavior and human relations, though the analogy may not as yet be fully realized.

The assumptions of the medical men and their allies are by now fairly well accepted in modern industrialized nations, though in practice the value placed on human life certainly is subject to wide variations. The assumptions essential for the proper functioning of the psychologist, anthropologist, and sociologist in our unique society, however, are, I believe, as special as the history of this society itself. The equivalent of the Hippocratic oath to which these men might well subscribe would therefore be related closely to the type of society in which they proposed to operate. Even the English and the American versions might vary at several points, but the essentials would be the same. Totalitarian nations, however, would use the techniques developed by these scientists for very different ends; and their use would condition the further advancement of the sciences themselves. Powerful tools are in the process of being forged by the scientists who study man as a social animal. These tools can be used to further or to destroy certain types of behavior and certain social patterns. Therefore it is essential for these men themselves to clarify in their own minds their own standards of value in many matters, just as, long ago, the medical profession settled certain issues confronting those whose knowledge includes the key to the life or death of an individual in distress.

SCIENTISTS AND THE GOVERNMENT

The traditions of science like the ethics of the medical profession are the product of a social organization international in scope and independent of any state. Can such independence be preserved today when the existence of an industrialized nation is so intimately associated with the applications of scientific knowledge? This is a serious question to which there can be no quick and reassuring answer. Those who look askance at any support deriving from public funds delight in dwelling gloomily on such a question. Yet with or without dollars from the taxpayer, the scientist today cannot escape from a situation where his specialty is a matter of vital consequence to organized society. Whether we like it or not, we live in a period when government must be increasingly aware of science and scientists of government. The politician and the research worker can no longer ignore one another. The relation of science to the modes of modern war alone assures a close connection between what were once two widely separated professions—that of the statesman and of the investigator of nature. Matters standing thus, we must conclude that the future health of science will be determined in no small measure by the action of the agents of the state. And in a democracy this means that public opinion will play a predominant part.

The continual vigor of research in pure science demands, in the United States, an electorate as informed as possible on all matters vital to the progress of research. The point has been already emphasized more than once in the preceding pages. But one more illustration of the interplay between science and society may be in order. No more vexing example of the type of problem that confronts both politicians and scientists today can be found than that of secrecy and censorship. Since the close of World War II there has been a great deal of discussion of the dilemma of the physicists and nuclear chemists who were in one way or another associated with the manufacture of atomic weapons. The designs of new military equipment had always been kept highly secret for reasons that are obvious; therefore, everything to do with the atomic bomb should be closely guarded, one may argue. Yet science can go forward only if there is no secrecy. Here is a real dilemma. Since the chain from the laboratory to the battlefield is continuous in the 1950's, who is to say what is science and therefore public, and what is basic to new military equipment and therefore secret? To state the problem is sufficient to show the inevitable tensions.

In spite of the fact that the heavy entanglement of science with weapon production is a product of the 1940's, the difficulty of reconciling science and secrecy is not new. At the close of World War I chemists employed by rubber manufacturers were not permitted to discuss even the fundamental chemistry of rubber with anyone outside the company. But a few years later all this was changed, and a Rubber Section of the American Chemical Society became an open forum for discussions of many basic problems. Patent protection to a considerable degree replaced the wall of complete concealment. In the heyday of German science, more than one eminent professor acted as consultant for a chemical firm; new processes for dyes or drugs discovered in his laboratory were patented with profit both to the company and to the professor. When such activities were in progress, the discussion of the day-to-day accomplishments was limited to a very few individuals; even the labels on the bottles might be in code. Yet in such cases the "blackout" was limited to a small group and was of relatively short duration. Nevertheless, that even such attempts at concealment can spoil the atmosphere of a laboratory has been attested to by many witnesses. The antithesis between science and secrecy is written in large letters on the record of history.

There are real dangers that the progress of science in the United States may be greatly impeded if the general public fails to understand the significance of free publication and discussion. Of course, one cannot expect those in charge of national defense in days of an armed truce to relax their vigilance; they are bound to be overconcerned with secrecy. That is why I believe it of importance that government support of pure science (apart from applied research and development) flow through a National Science Foundation. Though the forebearance of the Defense Establishment in these matters has been great and the farsightedness of the officers commendable, there is a fundamental incompatibility in having the progress of an international public enterprise closely related to preparation for national defense.

But is military secrecy the only obstacle to the advance of science that the grim postwar period has produced? Is it still true in 1950 that science is an international public enterprise? Unfortunately not. One of the tragedies of the divided world is the partisan, doctrinaire approach to intellectual and cultural activities which has been made increasingly explicit by the rulers of the Soviet Union and its satellites. Their attitude involves no denial of the importance of science; quite the contrary. More than one observer in the course of the last two decades has been impressed by the deep concern for science manifested by the Kremlin. American scientists who by special invitation attended the celebration of the Russian Academy of Science in the summer of 1945 noted with satisfaction the role played by Stalin in the public ceremonies honoring the scientists. One might jump to the conclusion that the interest in science shown by the highest officials of the Soviet Union proceeded solely from their realization of the importance of modern science to technology. While this realization plays its part, it would be a mistake to regard it as the only or even the prime factor.

In the *History of the Communist Party of the Soviet Union,* an official publication, the authors speak of the "tremendous part in the history of the party" played by a book of Lenin's, *Materialism and Empirio-Criticism,* published in 1909. This volume, the official historians of the Bolshevik party go on to say, "safeguarded a theoretical treasure from the motley crowd of revisionists and renegades." It is important to realize that the controversy in which the writing of the future ruler of Russia is alleged to have played so critical a role involved the nature of scientific truth.

Forty years and more ago, so we are told, the whole future of the Communist party was threatened by false doctrines concerning the validity and meaning of scientific principles in the field of physics. Can one wonder that a monolithic political party which so interprets its own history must continue to regard scientific theories and their interpretation as within the field of official competence? As the history to which I have just referred states clearly, no revolutionary party can accept any such doctrine as unity based on diversity; on the contrary, the success of the party has from the first been built on a rigid conformity enforced by the expulsion of those deviationists who did not understand or would not understand the "science of the development of society" as laid down in the Marxist-Lenin theory.

There is no indication that those in control of the party today are any less determined than their predecessors. Scientific theories that do not fit within the framework of the official version of dialectical materialism are clearly heresies; once convinced of this, the erstwhile proponents will of course publicly acknowledge their errors. Difficult as this may be for many in the western democracies to understand, the phenomenon of loyal sons of a church admitting their mistaken views and recanting is surely nothing new in history.

The fact that experimental findings in biology may be closely related to broad generalizations about heredity and thus have an impact on political and social theories obviously enhances interest in controversies in this particular field of science. But it is interesting to note that within the year *Pravda* has published at least one article devoted to a critique of modern theoretical physics; matters that would not incite even the passing interest of ninety-nine politicians out of a hundred in the western democracies are treated therein as of deep significance.

One may say that Communist edicts in genetics (p. 223) merely show that relatively ignorant and ruthless men are determining party policy in the field of science. But is it not inherent in any authoritarian system that human beings with all their fallibility must determine from year to year what is true doctrine and what is not? Politics, often of the crudest personal sort, will influence their decisions. The proponents of dialectical materialism in all lands place the physical sciences in a high position and speak glibly and confidently of the scientific method. When one version of this philosophy is transformed into the official doctrine of a party which must harbor no dissenters, the freedom of scientific thought automatically disappears. This is not to say that within wide limits scientific inquiries may not be ardently supported and that technology may not flourish. But can there ever be genuine scientific freedom in a society where all philosophical opinions must conform to the official interpretation of party dogma? This is a question which inevitably arises when one studies the history of the Communist party.

In an article on "Lenin and Philosophical Problems of Modern Physics," published in *Pravda* in May, 1949, S. I. Vavilov, the president of the Academy of Sciences of the U.S.S.R., described the problem of physics and politics in the following terms:

"Soviet physics, as well as all Soviet science, long ago entered the life of the state, having directed all its forces, in the service of our native land, toward satisfying the requirements of the great work of building communist society.

"Soviet physicists base their work on the world outlook of dialectical materialism, raised to a higher level by the genius-inspired works of Lenin and Stalin. But we cannot ignore the fact that some of our physicists still preserve idealistic survivals, supported chiefly by an uncritical reception of the literature of capitalist countries.

"It is our urgent task to fight these survivals by merciless criticism and self-criticism. Their harmful influence is great. Physicists must become more active in fighting them . . ."

Perhaps the most revealing statement of the orthodox Communist position has been given in an article published in May, 1950, in the English weekly, *Nature*. Here a member of the Institute of Genetics of the Academy of Science of Moscow replies to the criticism of Julian Huxley. Referring to Huxley's statement that "a great scientific nation has repudiated the universal and supranational character of science," the Russian scientist declares this is not true. "Soviet science never accepted these reactionary claims; it always combatted them," he states, and then continues, "We have openly declared and continue to declare that science, and therefore Soviet science, is a partisan science, a class science. . . . The bourgeoisie and its ideologists, whether they be biologists or not, were always afraid of openly avowing the partisan nature of science. . . . All the verbal nonsense about the 'supra class' and 'supranational' nature of science serves Huxley's absolutely definite class aims."

When we read such statements coming from Russian scientists in 1950 we seem to have stepped into another universe of discourse. Science as I have described it and defined it is not science in the eyes of a follower of the party line. All that scientists almost without exception take for granted in the free nations is repudiated and ridiculed on the other side of the Iron Curtain. Under these conditions there can be no real community of interests between the scientists of the two worlds; any cooperation or exchange of information will be merely a fortunate accident. In short, all science which conforms to the orders of the Executive Committee of the Communist party must be regarded as a new social phenomenon. Not only are scientific articles in Moscow subject to censorship in terms of military secrecy but the authors are under quite special social pressures.

Such being the case, should we discard the premise that the advance of knowledge is an international undertaking?

Not at all. For the time being we must reluctantly place the scientists on the other side of the Iron Curtain in a special category; unlike the rest of us they do not see in science an activity which knows no national boundaries. But having temporarily lost the allegiance of this group, it is all the more important that we ourselves stress the non-secret, international character of science. Even if the traffic in scientific discoveries and ideas were to become all one way (i.e., from the free nations to Russia) we would be wise to continue the tradition of unfettered international communication.

In the technological competition between business firms, the most progressive company has the most to gain by the advance of science. So the United States as a nation can benefit more than any other country by the continued vigor of scientific investigation. We are "tooled up" with men and equipment for applied research and engineering developments; we are ready to take the new theories and new discoveries as they come from the pure science laboratories and make the most of them. Therefore, whatever any other nation may do, whatever the tension of the times, we must continue to foster science; and that means fostering freedom of inquiry, of discussion, and of publication.

BIBLIOGRAPHY

The following list of books includes those mentioned in the preceding pages and certain others which may be of interest to the reader:

Armitage, Angus, *Sun, Stand Thou Still. The Life and Work of Copernicus the Astronomer*. New York, Henry Schuman, 1947.

Bates, Marston, *The Nature of Natural History*. New York, Charles Scribner's Sons, 1950.

Baxter, James Phinney, *Scientists Against Time*. Boston, Little, Brown and Company, 1946.

Bernal, J. D., *The Social Function of Science*. London, George Routledge and Sons, 1939.

Beveridge, W. I. B., *The Art of Scientific Investigation*. London, W. Heinemann, 1950.

Bridgman, P. W., *The Logic of Modern Physics*. New York, The Macmillan Company, 1927.

Bush, Vannevar, *Modern Arms and Free Men*. New York, Simon and Schuster, 1949.

Butterfield, Herbert, *The Origins of Modern Science*. London, G. Bell and Sons, 1949.

Clark, G. N., *Science and Social Welfare in the Age of Newton*. Oxford, 1937.

Cohen, I. Bernard, *Science, Servant of Man*. Boston, Little, Brown and Company, 1948.

Crowther, J. G., *Men of Science*. New York, W. W. Norton and

Company, 1936; *The Social Relations of Science*. New York, The Macmillan Company, 1941.

Dewey, John, *Logic, The Theory of Inquiry*. New York, Henry Holt and Company, 1938.

Dickinson, H. W., *James Watt, Craftsman and Engineer*. Cambridge University Press, 1936.

Dubos, René J., *Louis Pasteur, Free Lance of Science*. Boston, Little, Brown and Company, 1950.

Frank, Philipp, *Einstein, His Life and Times*. New York, Alfred A. Knopf, 1947.

French, Sidney J., *Torch and Crucible. The Life and Death of Antoine Lavoisier*. Princeton University Press, 1941.

Geikie, Sir Archibald, *The Founders of Geology*. London and New York, The Macmillan Company, 1905.

Harvard Case Histories in Experimental Science. Cambridge, Harvard University Press, 1950:

Case 1. *Robert Boyle's Experiments in Pneumatics*. Edited by J. B. Conant.

Case 2. *The Overthrow of the Phlogiston Theory*. Edited by J. B. Conant.

Case 3. *The Early Development of the Concepts of Temperature and Heat*. Prepared by Duane Roller.

Case 4. *The Atomic-Molecular Theory*. By Leonard K. Nash.

Howard, Frank A., *Buna Rubber, the Birth of an Industry*. New York, Van Nostrand Company, 1947.

Huxley, Julian, *Heredity, East and West*. New York, Henry Schuman, 1949; *Evolution, the Modern Synthesis*. New York, Harper Brothers, 1942.

James, William, *The Philosophy of William James. Drawn from His Own Works*. Introduction by H. M. Kallen. New York, The Modern Library, 1925.

Killeffer, D. H., *The Genius of Industrial Research*. New York, Reinhold Publishing Corporation, 1948.

Maclaurin, W. Rupert, *Invention and Innovation in the Radio Industry*. New York, The Macmillan Company, 1949.

Mees, C. E. K., and Leermakers, J. A., *The Organization of Industrial Scientific Research*. New York, McGraw-Hill Book Company, 1950.

Pepper, Stephen C., *World Hypotheses*. Berkeley, University of California Press, 1942.

Polanyi, M., *The Contempt of Freedom*. London, Watts and Company, 1940.

Singer, Charles, *A Short History of Biology*. Oxford, 1931.

Stimson, Dorothy, *Scientists and Amateurs. A History of the Royal Society*. New York, Henry Schuman, 1948.

White, A. D., *A History of the Warfare of Science and Theology*. New York, D. Appleton and Company, 1897.

Whitehead, A. N., *Science and the Modern World* (Lowell Lectures, 1925). Pelican Mentor Edition, 1948.

Whittaker, Sir Edmund, *From Euclid to Eddington*. Cambridge University Press, 1949.

Wightman, W. P. D., *The Growth of Scientific Ideas*. New Haven, Yale University Press, 1951.

INDEX

Abstract ideas, 162

Académie des Sciences. *See* French academy

Accademia dei Lincei, 18

Accademia del Cimento, 18–19, 78, 84, 85, 100, 103, 153, 212, 232, 233

Accident, role of, in the advance of science, 108–121 *passim*

Accidental errors, 180–181; *see also* Experimental errors

Accumulative knowledge, 37–38, 39, 40, 263

Agricola, 34, 67, 68

Agriculture, 34, 208, 215; scientific knowledge applied to, 313

Agriculturists, 40, 46, 58, 313

Air, composition of, 116; compressibility of, 94–95, 138, 142–143; density of, at varying altitudes, 147–149; and the phlogiston theory, 170–171

Air pump. *See* Pumps

Alcoholic fermentation. *See* Fermentation

Alleged scientific method, 7–8, 42–43, 45, 50, 197

Almond oil, 227, 229

Altimeter, barometric, 150

Altitude, relation between atmos-

pheric pressure and, 70–71, 72–73, 75, 147–151 *passim*

Amateur, contributions of, to advance of science, 17, 77–78, 79, 284, 316, 318

American citizen, and the importance of understanding science, 1–6, 346–347

American Revolution, 166, 169

Ammonia, 117

Animal electricity, and Galvani's discoveries, 109–112; question of existence of, 113–114

Anthropologist, 342

Anthropology, 37, 41; *see also* Social sciences

Appert, François, 244

Applied research, goals of, 305; modern trend toward, 303, 319, 320

Applied science, dependent of, upon pure science, 326; in industry, 61; and medicine, 315; nature of, 59–60; versus pure science, 298, 302–304

Aquinas, Thomas, 38

Archaeologists, 44

Archaeology, 37; in relation to science and history, 294–295

359